T0332388

CARBON NANOTUBES

Angels or Demons?

SILVANA FIORITO

University "La Sapienza" – Institute of Neurobiology and Molecular Medicine, CNR-Rome, Italy

CARBON NANOTUBES

Angels or Demons?

PAN STANFORD PUBLISHING

Published by

Pan Stanford Publishing Pte. Ltd.
5 Toh Tuck Link
Singapore 596224

Distributed by

World Scientific Publishing Co. Pte. Ltd.
5 Toh Tuck Link, Singapore 596224
USA office: 27 Warren Street, Suite 401-402, Hackensack, NJ 07601
UK office: 57 Shelton Street, Covent Garden, London WC2H 9HE

British Library Cataloguing-in-Publication Data
A catalogue record for this book is available from the British Library.

CARBON NANOTUBES: ANGELS OR DEMONS?

ISBN-13 978-981-4241-01-4
ISBN-10 981-4241-01-6

Typeset by Stallion Press
Email: enquiries@stallionpress.com

Printed in Singapore by B & Jo Enterprise Pte Ltd.,
E-mail: sankaran@rpsonline.com.sg

To my Parents

Preface

Nanotechnology can be regarded as the major technological challenge of this century that is stirring people's imagination about its potential use. But "Nano" should also be viewed as the major intellectual, social and moral challenge of this century — as an exquisite field for testing our ideas about how people generate knowledge, how they integrate new technologies into their practices and organize themselves around new kinds of artifacts, how they use emerging technologies to push the limits of human instrumentality. A new era has begun that is likely to change people's way of life, thinking and behaving in a very deep manner.

Thinking "small" is a revolutionary tool for making people's mind to be more focused on the "small" rather than on the "big", on the "inside" rather than on the "outside", on the details rather than on the showy aspects of the reality. Thinking "small" can influence people's ethical and social behaviour in terms of how they interact with each other, how they express their feelings, analyse and interpret physical, natural, social, historical, artistic phenomenon. Thinking "small" can represent the starting point for the reversal of the trend whose worse aspects lead some people to pursue ideals of greatness, to get the biggest house, car, fridge, television etc. in order to parade their own grandeur. Thinking "small" can stimulate minds to be more analytical, to look "behind" the appearances instead of at the surface, to swim deeply under water exploring a mervellous hidden world instead of floating superficially.

Nanoscientists manipulate objects and forces at a scale of one millionth of a hairs width. At that size, matter behaves differently. Light and electricity resolve into individual photons and electrons, particles pop in and out of existence, and other once-theoretical oddities of quantum mechanics are seen to be real. Nanoscale research encompasses communications, new materials, and the study of life, as well as weird quantum phenomena and incidental things that exist in the real world, like viruses, dust and diesel exhaust. Physics, mechanical and electrical engineering, materials science, chemistry, biology, and medicine converge here. This is the realm of the lowest common denominator.

Since nanotechnology was recognized as "the technology of the 21st century" billions of dollars of funding have been invested in the USA, Europe and Japan and a lot of investigations are being in progress about how this technology might affect our future. The National Science Foundation predicts that the market for nanotech products will reach $1 trillion by 2015. "We have learned that thinking very small may be even bigger than thinking big," said Mark Tyler, president of a company, which had about $25 million in sales last year. But the vast majority of people still have little or no idea of what nanotechnology is or what the possible implications of this technology might be.

There is a high level of enthusiasm for the potential benefits of nanotechnology and little concern about its possible dangers. The general public perception of nanotechnology is still viewed in a favourable light, and public mentally connects it with other areas of science, such as the space program, nuclear power and cloning research, that are thought to be as well as a kind of science that is between science fiction and reality and that, for this reason, can only lead to positive and useful results for the human, animal and environmental well being in the next future. The public anticipates that nanotechnology will advance environmental protection, lower energy costs, and provide better food and nutrition products, according to a report from the Woodrow Wilson International Center for Scholars and the Pew Charitable Trusts.

Nano-supporters say that nature (or 'biology') has been doing nanotechnology for billions of years; every virus, bacterium, and cell is a nanomachine of enormous complexity. Nature's nano-achievements show us that nanomachines are possible, and nature's version of nano has completely restructured the earth and produced human life, culture, and consciousness. The progress of science, they say, means that it will inevitably be possible for us to understand and mimic nature's nanomachines; once we have done so, our own nanomachines will develop in a way determined by biology, chemistry, and engineering design; and as they do develop, our inventions cannot help but revolutionize our world just as much as nature's nanorobots did (Mody Cyrus C.M., 2004). Eric Drexler, a writer who can be regarded as the popularizer of the term 'nanotechnology' and one of the most influential visionaries of the field, by 1981 begun publishing his vision under the label of 'nanotechnology' — a vision in which very small 'assemblers', modelled on biological machines (cells, ribosomes, viruses, etc.), could reconstitute raw materials into almost any physically possible artefact. He tends to focus on the very ancient *biological* precursors of nanotechnology, since this helps to make the analogy between biological and artificial nanomachines (Drexler K.E., 1981). Others, outside the Drexler camp, are more likely to point out very old *craft* activities that would today count as 'nanotechnology': the process of nanofabrication, in particular the making of gold nanodots, is not new. Much of the color in the stained glass windows found in medieval and Victorian churches and some of the glazes found in ancient pottery, depend on the fact that nanoscale properties of materials are different from macroscale properties. In some senses, the first nanotechnologists were actually glass workers in medieval forges rather than the bunny-suited workers in a modern semiconductor plant. Clearly the glaziers did not understand why what they did to gold produced the colors it did, but we do understand now (Ratner M. & Ratner D., 2003). Nano, in this formulation, produces new *knowledge* that maps onto old *practice*. What makes nano new is that it brings *understanding* where before there was only *doing* (Mody Cyrus C.M., 2004) The nanoscale has become a place that tourists can visit, where everything is different, yet exactly the same — all the building blocks are atoms, at which we should wonder, but they are being used to make ordinary, familiar, everyday objects, whose use is something we intuit rather than theorize about. Nevertheless, this view could have positive as well as negative effects on people's minds. For instance, discoveries and developments within nanotechnology research and especially within carbon based nanotechnology (Carbon nanotubes and fullerenes) have led to some peculiar but also serious projects on various applications. Best known and widely referenced is a project called "the space elevator" from NASA, a 62,000-mile twine of carbon nanotubes that would transport cargo into orbit. This system is decades away from being feasible, nevertheless the public acceptance of this project is quite high as it is something that everybody can relate to in the sense that everybody knows an elevator and everybody knows that we are able to travel in space. Thus, it has the positive side that it can provide the general public with a gentle introduction to nanotechnology developments and allow them to learn more about carbon nanotechnology. On the other hand, the negative view is that this project is almost impossible to achieve and it gives the audience a wrong view of what to expect from nanotechnology in the near future.

The effects of future, yet unrealized technologies are in most cases subject to great uncertainty. Nanotechnology is an unusually clear example of this. As already mentioned, the technological feasibility of the nanoconstructions under ethical debate is in most cases uncertain. Furthermore, many of the possible future nanotechnologies are so different from previous technologies that historical experience provides very little guidance in judging how people will react to them. The development and use of new technologies is largely determined by human reactions to them, these have their influence via mechanisms including markets, politics and social conventions (Rosenberg N., 1995). It is not only the negative but also the positive effects of nanotechnology and other future technologies that are subject to great uncertainty. The most fervent proponents of nanotechnology have argued that it can solve many of humanity's most pressing problems: Nanotechnology can make cheap solar energy available, thus solving the energy problem. Nanoscale devices injected into the bloodstream can be used to attack cancer cells or arterial plaques, thus eradicating major diseases. Synthetic human organs can be constructed that replace defective ones.

Much of the public discussion about nanotechnology concerns possible risks associated with the future development of that technology. But analyzing technological risks, namely risk analysis, for guidance about nanotechnology is quite difficult. The reason for this is that the tools of risk analysis have been tailored to deal with other types of issues than those presently encountered in connection with nanotechnology. Risk analysis was developed as a means to evaluate well-defined dangers associated with well-known technologies. But nobody knows today whether or not any of these types of nanodevices will ever be technologically feasible. Neither do we know what these hypothetical technologies will look like in case they will be realized. Therefore, discussions on such dangers differ radically from how risk analysis is conducted. The tools developed in that discipline cannot be used when so little is known about that kind of nano-objects.

Carbon-nanotubes based nanostructures (C-nanotubes) in particular are viewed as a class of nanomaterials with high potential for biological applications due to their unique mechanical, physical and chemical properties. Among numerous potential applications, including DNA and protein sensors, *in vitro* cell markers, diagnostic imaging contrast agents, their use as multifunctional biological transporters, agents for selective cancer destruction and drug and gene delivery systems has been explored. Moreover, various cell types have been shown to grow extremely well on C-nanotubes, giving a potential for applications such as scaffolds and structures/coatings for tissue regeneration/repair. Furthermore C-nanotubes reinforced composites demonstrated a substantial promise for orthopaedic and dental devices as well as vascular stents.

Before medical applications can be developed it is necessary to explore the behaviour and fate of engineered C-nanostructures in mammals and environment. However, actually little is known in this area and results are often contradictory and not conclusive yet, in part because of the challenge of detecting and tracking these nanoparticles in complex biological environments, in part due to the different factors influencing their toxicity (length, degree of purity, presence of metal catalysts, degree of aggregation, functionalisation, etc.).

Not necessarily the real risks of these nanomaterials are high, but the uncertainty of their impact on health, safety and environment are a serious threat to further progress in nanosciences and even more so in nanotechnologies.

The purpose of this book is to provide an overview on the expected benefits and the possible risks of carbon-based nanotechnology and to evaluate the potential health, safety and environmental implications of carbon-based nanostructures in order to make people aware of what is going to change their life in the next future.

Silvana Fiorito

Contents

Preface vii

1. **Carbon Nanotubes: A Basic Description** 1

Christophe Goze-Bac

1.1. Brief History of the Nanotubes 1
1.2. Size and Shape of Nanotubes 2
1.3. Synthesis of Nanotubes 3
1.4. Molecular Structures of Nanotubes 5
1.5. Electronic Properties of Carbon Nanotubes 12
1.6. Recent Works on Individual Carbon Nanotubes 13
Bibliography 15

2. **Applications in Mechanics and Sensors** 19

Angel Rubio

2.1. Introduction 19
2.1.1. Some Applications: Carbon Nanotube-Based Sensors 21
2.2. Nanotube Nanomechanics 22
2.2.1. Experimental Evidence of Nanotube Resilience . 26
2.3. Applications of Mechanical Response 27
Acknowledgements 30
Bibliography 30

3. **Applications in Biology and Medicine** 35

Guido Rasi and Claudia Matteucci

3.1. Application in Biosensing, Imaging and Tissue Engineering 36
3.2. Application in Drug, Peptide and Gene Delivery 46
3.3. Application in Cancer Diagnosis and Therapy 52
3.4. Conclusions 56
Bibliography 57

4. **Carbon Nanotubes: Applications for Medical Devices** **61**

Robert Streicher

4.1. Introduction 61
4.2. Biomaterials and Implants 62
4.2.1. Orthopaedic Implants 63
4.2.1.1. Current Issues with Orthopaedic Implants 64
4.3. Nanomaterials in Medicine 67
4.4. Carbon Nanotubes 67
4.4.1. Biomedical Applications 68
4.4.2. Interaction with Biologic Structures 69
4.4.2.1. Tissue Engineering for Neural Applications 71
4.4.2.2. Tissue Engineering for Orthopaedic Applications 77
4.4.3. Structural Components for Orthopaedics 81
4.4.3.1. Carbon Fibers and Composites for Orthopaedic Applications 82
4.4.3.1.1. Polymer/CNT Nano-composites 86
4.4.3.1.2. Alumina/CNT Nano-composites 90
4.5. Summary and Conclusions 91
Bibliography 94

5. **Toxicity of Carbon-Nanotubes** **105**

Silvana Fiorito

5.1. Behaviour and Fate of Carbon-Nanotubes in Mammals . 106
5.2. Cellular Uptake of Carbon-Nanotubes 108
5.3. Lung Toxicity of Carbon-Nanotubes 108
5.4. Cell Toxicity of Carbon-Nanotubes 113
Acknowledgement 122
Bibliography 122

6. **Mechanisms of Toxicity of Carbon Nanotubes** **127**

Silvana Fiorito and Annalucia Serafino

6.1. Cytotoxicity Mediated by Oxidative Stress 127
6.2. Cytotoxicity Mediated by Altered Gene Expression . . . 129

6.3. Cytotoxicity Mediated by Geometrical and Mechanical Factors 131
6.4. Cytotoxicity Mediated by the Binding of CNTs to Plasma Proteins 132
Acknowledgement 133
Bibliography 133

7. Conclusions **137**

Index **139**

1 Carbon Nanotubes: A Basic Description

Christophe Goze-Bac

1.1. Brief History of the Nanotubes

The experiments on the synthesis of new carbon clusters during the mid-1980s by Harry Kroto and Richard Smalley are the prelude to the story of nanotubes. From the vaporization of graphite and its expected condensation in carbon clusters of different sizes, particular structures containing precisely 60 and 70 carbons appear to be much more stable. These unexpected results brought into the spotlight the now famous C_{60} and its closed shell similar to a soccer ball, as presented in Figure 1. The family of the fullerene is discovered and published in *Nature* in 1985.[1] Five years later, the success in the synthesis of C_{60} in bulk quantity by Wolfgang Kratschmer and Donald Huffman[2] gave to the community, the opportunity to study the properties of this new form of carbon. During the 15 years, an impressive number of publications and extraordinary results appeared on the bucky ball and its derivatives.[3,4]

During this time in the early-1990s, Sumio Iijima was investigating by electron microscopy technique the carbon soot and deposit produced by a Kratschmer-Huffman's machine. From his findings, he concluded to the existence of novel graphitic structures, another new form of carbon similar to tiny tubules of graphite presumably closed at each ends.[5] Figure 2 shows a high resolution transmission microscopy image of one multi-walled carbon nanotubes produced by the electric arc method very similar to the hollow carbon cylinders observed in 1991. From historical point of view, if the anterior existence of nanotubes is well established,[6–8] Iijima's work pointed out to the exceptional properties and potential applications of these nanostructures. An important step was accomplished in 1992, by improving the method of synthesis and the capacity of production with gram quantities.[9] In 1993, another key point is the successful synthesis of single walled carbon nanotubes which are considered to be the ideal nanotubes with a very single layer.[10–12] The diameter appeared to be amazingly small compared to multi-walled carbon nanotubes, as it can be seen between Figures 2 and 3. In 1997, high yields of single walled nanotubes were achieved by electric arc techniques with the help of optimized conditions.[13] This definitely extends the possibilities of investigations of single and multi-walled carbon nanotubes.

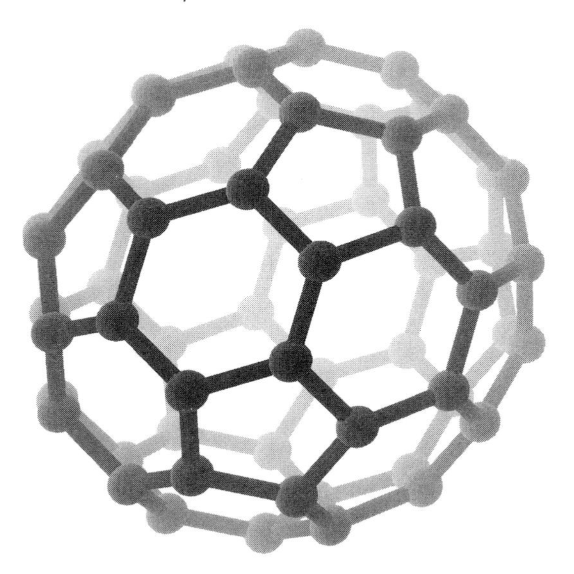

Fig. 1. The C_{60} molecule : the bucky ball.

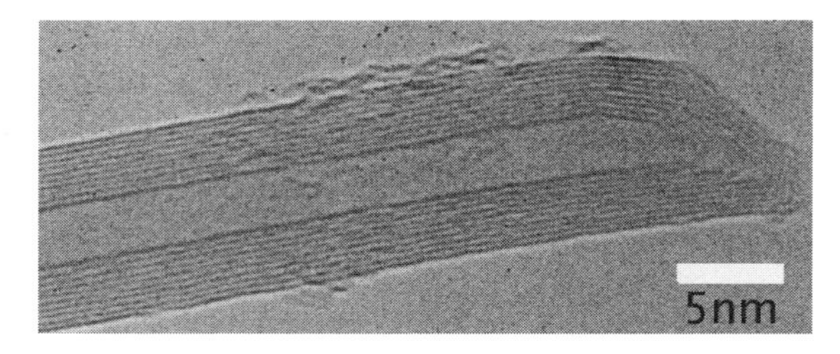

Fig. 2. High resolution transmission microscopy of MWNT produced by the electric arc method.

1.2. Size and Shape of Nanotubes

Downsized 10 000 times a human air, you will get the size and shape of a typical nanotube. These molecules are few nanometers in diameter and several microns in length. Single carbon nanotubes come often as tightly bundles of single walled nanotubes entangled as curly locks, see Figure 4. The packing of the nanotubes inside a bundle is more or less triangular with an intertube distance about 3.14 Å,[14,15] as shown in Figure 3. Here, the bundle is about 20 nanotubes but can be made of hundreds of individuals. This assembly of

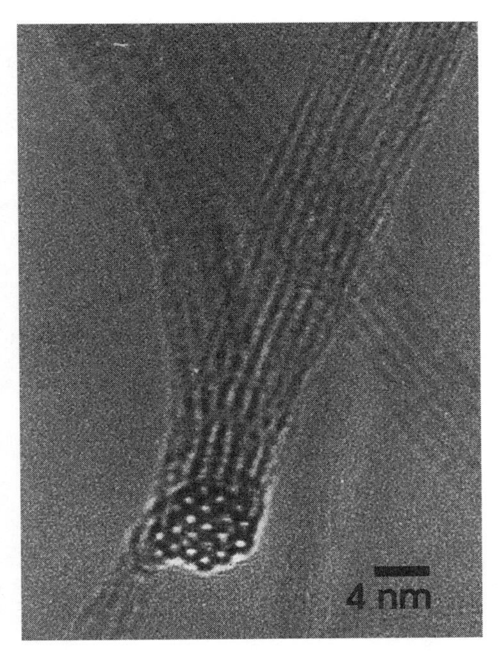

Fig. 3. High resolution transmission microscopy image of one bundle of single walled nanotubes.[13]

nanotubes is stabilized by van der Waals interactions. If the distribution of nanotube diameters is quite narrow within a bundle, it has been shown that the arrangement of the carbon atoms is diverse.[16,17]

In the case of multi-walled nanotubes, at least two concentric nanotubes of different diameters are encapsulated like Matrioschka Russian doll.[5] Their lengths are from hundreds of nanometers to tens of micrometers, with diameter of few to hundreds of nanometers.

1.3. Synthesis of Nanotubes

A wide variety of processes is now available to produce carbon nanotubes. They can be classified, first into high temperature methods like the electric arc,[2,5,9,13] laser ablation[12] and solar beam[18] based on the sublimation of a graphitic rod in inert atmosphere and second into process working at moderate temperature like the catalytic chemical vapor deposition which produces nanotubes from the pyrolysis of hydrocarbons.[19,20] Originally, the electric arc

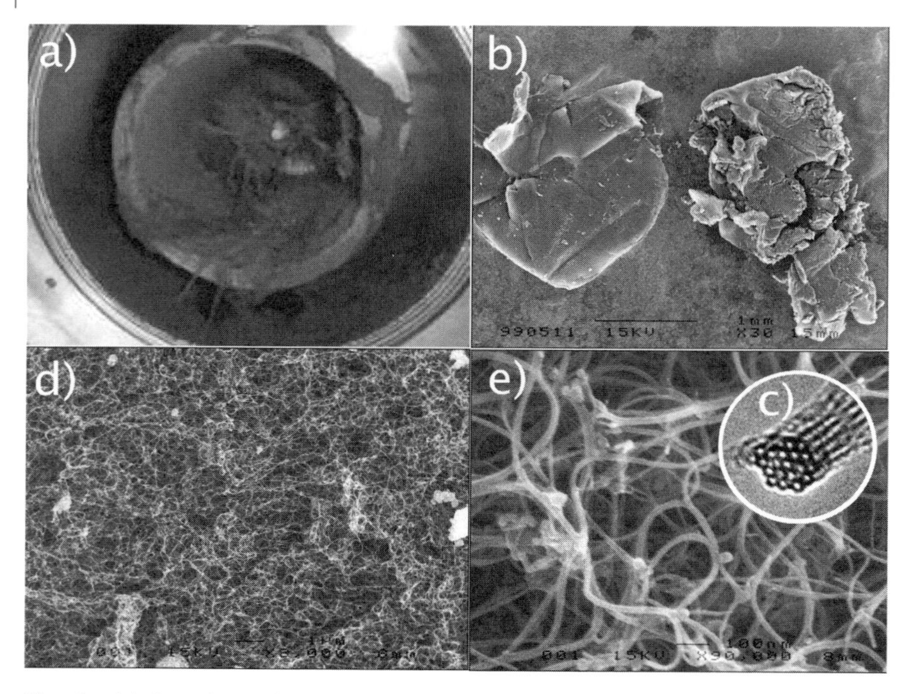

Fig. 4. (a) Open in air electric arc discharge reactor used to produce carbon nanotubes and fullerenes in Montpellier, France. Scanning electron microscographs of as-produced (b) millimetric grains collected from the central part of the collaret around the deposit. Entangled network of bundles of single walled carbon nanotubes at higher magnification (d) ×8000 and (e) ×90 000. The inset (c) presents a high resolution transmission micrograph of one bundle.

discharge setup developed by Kratschmer-Huffamn was invented for fullerene production. Such electric arc chamber is presented in Figure 4 (a). In a helium or argon atmosphere a DC current (100 amps 27 volts) is applied through two high purity graphite electrodes. In these conditions, the temperature can reach 6000 K which is high enough to sublimate continuously the graphite from the rod of the anode. At the cathode where the temperature is lower, a deposit and soot are produced. Fullerene like C_{60} and C_{70} are generally found in large quantities in the soot whereas nanotubes are present in the deposit and its vicinity. Later, this apparatus was successfully used by different groups to grow in a very simple and cheap way single walled nanotubes with the help of suitable catalyst like Ni, Co, Pt, Rh, Fe ... incorporated inside the anode.[13] In addition, it was rapidly shown that the characteristics of the nanotubes could be easily change by playing with the current, the pressure, the nature of the catalyst. This finding was a breakthrough and allowed the community to investigate the properties of carbon nanotubes to a very large extend. The picture of the electric chamber open in air presented in Figure 4 (a), is taken after running one synthesis. The reactor contains a lot of black soot, carbon filaments and the deposit. The material to be collected which contains a high density of carbon nanotubes is in the central part of the collaret and looks like webs. High temperature routes are known to produce high quality materials which mean higher graphitization of the nanotubes. They can be used also to produce large quantities of single walled carbon nanotubes. The as-produced nanotubes are presented at different magnifications in Figures 4 (b), (d) and (e) using scanning electron microscopy images. The inset (c) shows a high resolution transmission electron microscopy image of one bundle containing about 20 single walled nanotubes. We turn now to the chemical catalytic vapor desposition method. This process was originally developed in order to decompose hydrocarbons e.g. methane, acetylene, methane etc. over catalysts e.g. Ni, Co, Fe, Pt.[19,20] This technique allows to grow nanotube at predefined locations on substrates where nanosized metal particles are deposited. Typical transmission electron micrograph of MWNT is shown in Figure 5.

To finish, one has to mention that if some methods of production are now well established, nanotubes are still not prepared in pure form. For example, single walled carbon nanotubes samples are contaminated by graphitic structures, amorphous carbon and magnetic particles used as catalyst. In order to study the intrinsic properties of the nanotube, more or less sophisticated strategies of purification have been developed.[21,22] Only with the help of magnetic fields[23] used to trap magnetic impurities, highly purified bulk samples have been obtained which offer the opportunity to use magnetic resonance spectroscopy.[23,24]

1.4. Molecular Structures of Nanotubes

At a first glance, the molecular structures of carbon nanotube can be seen as single or multiple graphitic layers rolled to form a tubule. In the case of single walled nanotube, one atom thick layer is bended and welded as presented in Figure 6. Hence like in the graphite, the bonding between carbon atoms involves three neighbors and their hybridization are expected to be sp^2 like with a small *s* character because of the curvature. If this scenario is attractive and apparently simple, the way to form a nanotube is not unique and presents interesting features. Hence, it is challenging to classify these nanostructures

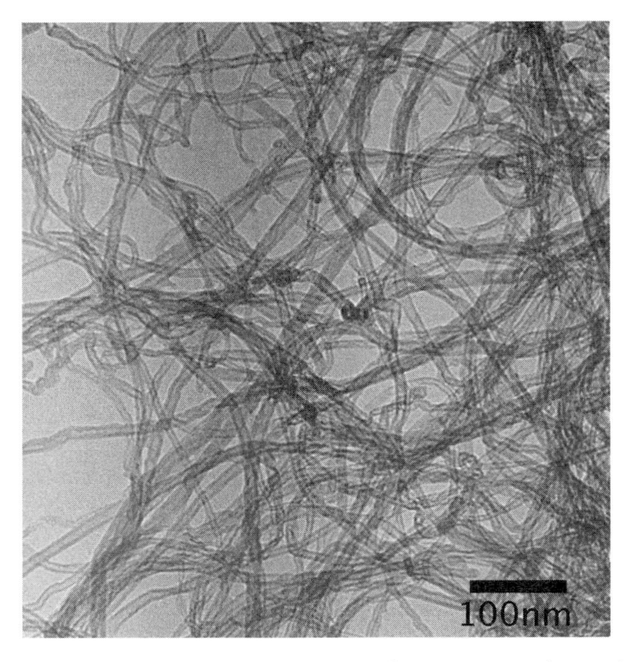

Fig. 5. Transmission electron microscopy images of MWNT grown by catalytic chemical vapor deposition methods.

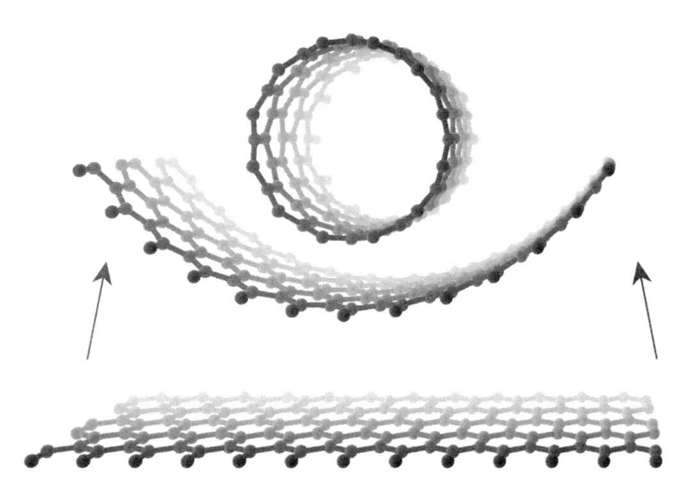

Fig. 6. Bending a one carbon atom thick graphitic layer in order to form a (12,0) carbon nanotube.

and to guess their properties. The high symmetry structures of nanotubes are generally divided in three types : armchair, zigzag and chiral as illustrated in Figures 7, 8 and 9, respectively. These labels come from the distribution of the bonds along the perimeter of the nanotube describing zigzag or an armchair. The chiral family presents forms of lower symmetries with hexagons arranged helically along the axis of the tube. Note that these three nanotubes have very similar radius $R \approx 6.6 \pm 0.4$ Å. Theoretical models of nanotubes have been

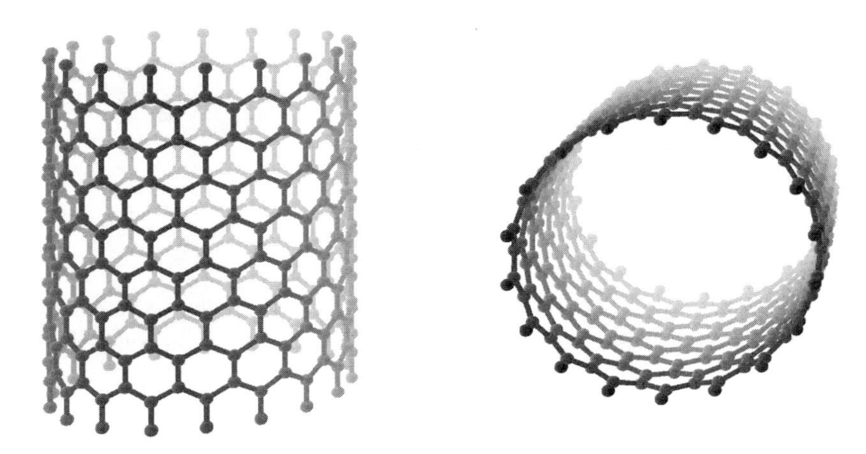

Fig. 7. A zigzag $(18, 0)$ nanotube.

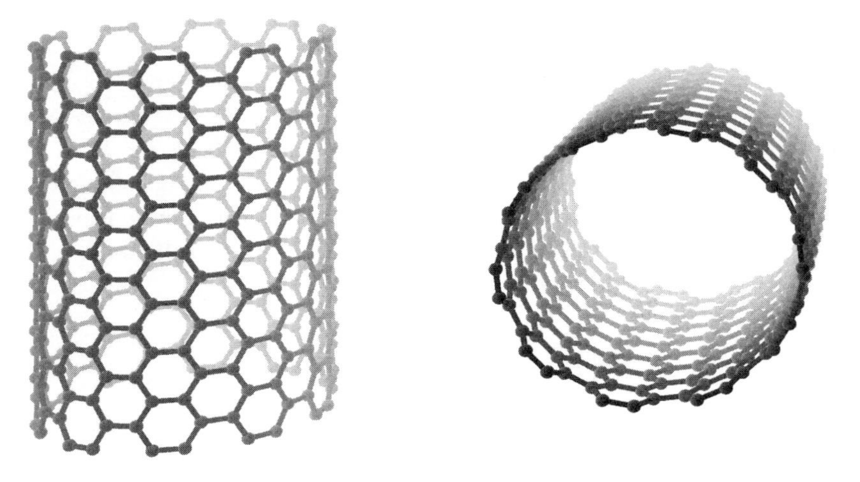

Fig. 8. An armchair $(10, 10)$ nanotube.

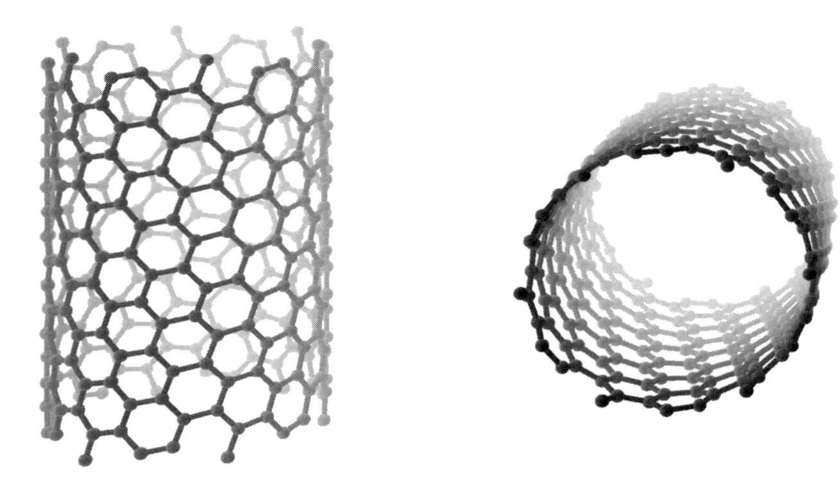

Fig. 9. A chiral (12, 6) nanotube.

proposed and a standardization is now available which helps to understand the main characteristics of both single walled nanotubes and multi-walled nanotubes.[3]

A. Single walled nanotubes: the vector model

One way to formalize the structures of single walled nanotubes is to use the vector model presented in Figure 10 which completely determines the structure of a particular nanotube with a pair of integer (n, m). The vector P is defined by joining two equivalent carbon atoms of the graphene layer. The nanotube is obtained by bending up P from head to tail giving its perimeter $|P| = 2\pi R$, where R is the radius of the nanotube. Before rolling up the nanotube, P is inside the graphene plane and can be read as a linear combination of a_1 and a_2 which are the base vectors of the unit cell of the graphene. With the convention $n \geq m$, we receive

$$P = n \cdot a_1 + m \cdot a_2 \tag{1}$$

It can be seen from Figure 10, that in cartesian coordinates

$$a_1 = \frac{a}{2} \cdot \begin{pmatrix} \sqrt{3} \\ -1 \end{pmatrix}; \quad a_2 = \frac{a}{2} \cdot \begin{pmatrix} \sqrt{3} \\ 1 \end{pmatrix} \tag{2}$$

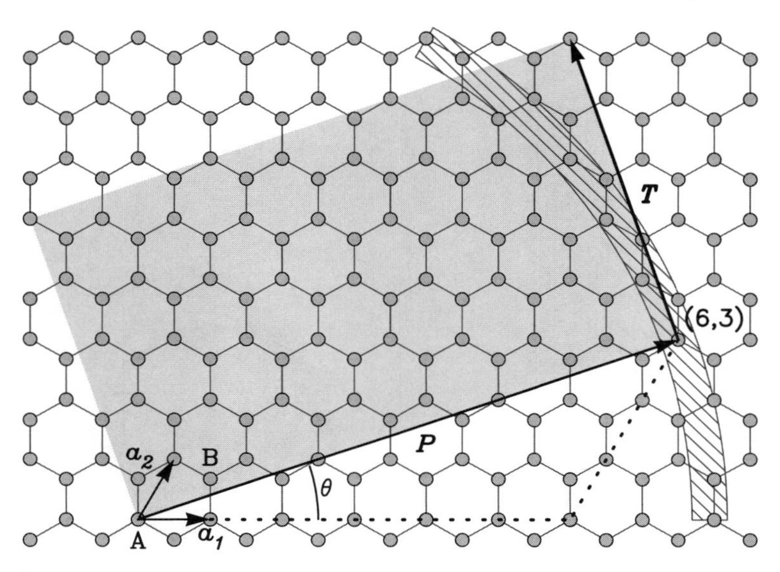

Fig. 10. Graphene layer with the two lattice vectors a_1 and a_2. Carbons A and B are the two inequivalent atoms. The perimeter vector P, the translational vector unit T, the chiral angle θ and the unit cell (grey area) are represented for the construction of the $(6, 3)$ nanotube. Note that if the head of the vector P is inside the dashed area, the corresponding nanotubes have similar diameters and present different symetries and geometries.

where $a = 2.461$ Å. The graphene has two inequivalent carbon atoms A and B per unit cell with the following coordinates

$$A : \begin{pmatrix} 0 \\ 0 \end{pmatrix}; \quad B : \begin{pmatrix} \frac{2a}{\sqrt{3}} \\ 1 \end{pmatrix} \tag{3}$$

Zigzag nanotubes correspond to P colinear to a_1 i.e. $m = 0$ and for armchair nanotubes $m = n$, i.e. $P = m \cdot (a_1 + a_2)$. In all the other cases, a chiral nanotube is constructed.

As a tutorial example, the vector $P = 6 \cdot a_1 + 3 \cdot a_2$ is presented in Figure 10. All the nanotubes constructed from carbon atoms inside the dashed area show similar radius than the $(6, 3)$. After rolling up the graphene sheet according to P, the chiral nanotube $(6, 3)$ is obtained as presented in Figure 11.

More generally, one can write the vector

$$P = \frac{a}{2} \cdot \begin{pmatrix} \sqrt{3} \cdot (n + m) \\ m - n \end{pmatrix} \tag{4}$$

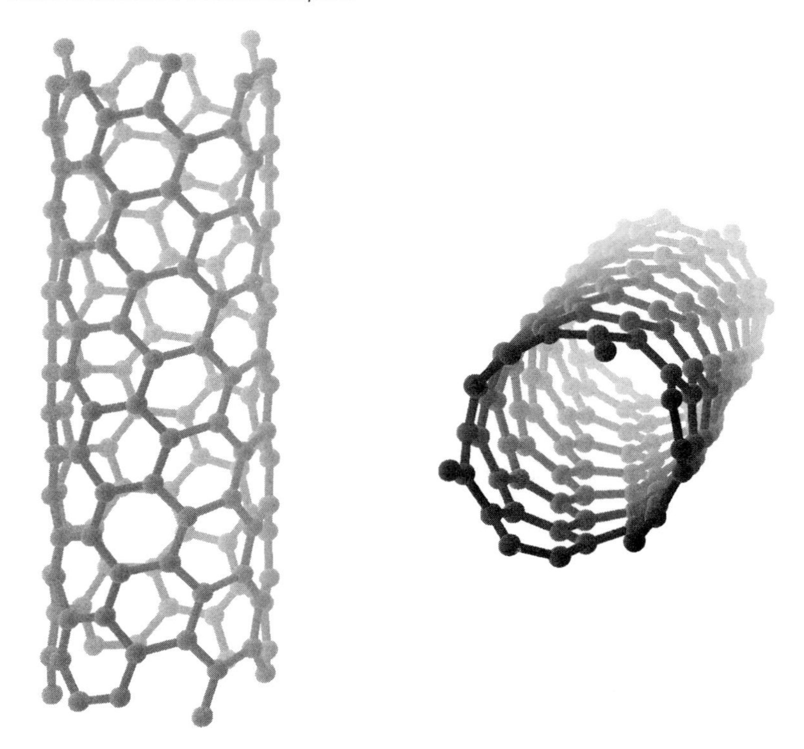

Fig. 11. A chiral $(6,3)$ nanotube corresponding to the rolled up graphene sheet displayed in Figure 11.

since $|a_1| = |a_2| = 2.461$ Å, the perimeter $|P|$ of a (n, m) nanotube is given by

$$|P| = |n \cdot a_1 + m \cdot a_2| = \frac{a}{2} \cdot \sqrt{(3(n+m)^2 + (m-n)^2} \tag{5}$$

then

$$|P| = 2.461 \cdot \sqrt{n^2 + m^2 + mn} \quad (\text{Å}) \tag{6}$$

which gives a radius for the nanotube

$$R = \frac{2.461}{2\pi} \cdot \sqrt{n^2 + m^2 + mn} \quad (\text{Å}) \tag{7}$$

Similarly, the chiral angle θ is written

$$\theta = \arctan \frac{\sqrt{3}m}{2 \cdot n + m} \tag{8}$$

A second vector T has to be defined which corresponds to the translational unit cell along the nanotube axis. The vector T for the $(6, 3)$ and its translational unit are displayed in Figure 10. It can be noticed that for zigzag and armchair nanotubes, the length of the unit cell is equal to $\sqrt{3} \cdot a$ and a, respectively. For chiral nanotubes it can be longer. This vector presents the two following properties :

- T is perpendicular to $P \Longleftrightarrow T \cdot P = 0$
- T is a linear combination of a_1 and $a_2 \Longleftrightarrow T = m' \cdot a_1 + n' \cdot a_2$ where m' and n' are integers

Thus, we have to solve the following equation

$$T \cdot P = \frac{a^2}{4} \cdot \begin{pmatrix} \sqrt{3} \cdot (n+m) \\ m-n \end{pmatrix} \cdot \begin{pmatrix} \sqrt{3} \cdot (n'+m') \\ m'-n' \end{pmatrix} = 0 \tag{9}$$

$$3 \cdot (n+m)(n'+m') + (m-n)(m'-n') = 0 \tag{10}$$

which gives

$$m' = -n' \frac{2n+m}{n+2m} \tag{11}$$

Here, we introduce d the greatest common divisor of $2n+m$ and $n+2m$, and we can rewrite

$$T = \frac{2m+n}{d} \cdot a_1 - \frac{2n+m}{d} a_2 \tag{12}$$

hence

$$T = \frac{a}{2d} \cdot \begin{pmatrix} \sqrt{3} \cdot (m-n) \\ -3(m+n) \end{pmatrix} \tag{13}$$

where

$$d = gcd(2n+m, n+2m) \tag{14}$$

A single walled nanotube has a one dimensional structure defined by the unit cell vector T. Its length is $T = \sqrt{3}|P|/d$. It can be show that one unit cell contains $4(n^2 + m^2 + nm)/d$ atoms, which can be large for chiral nanotube.

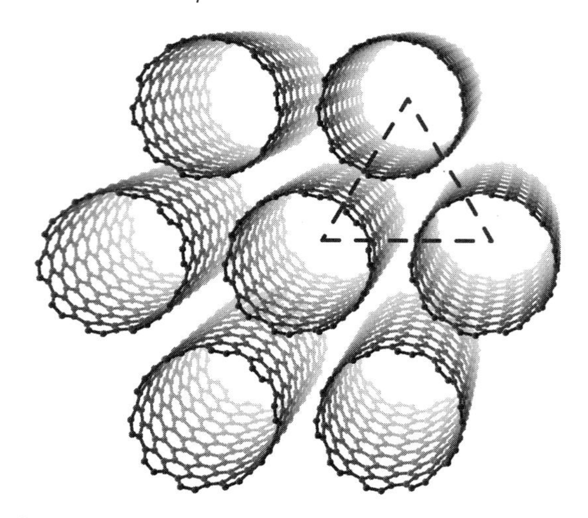

Fig. 12. A bundle of $(10, 10)$ carbon nanotubes. A more or less triangular lattice is generally observed experimentally. Note that this situation is far from reality since randomly distributed chiralities are experimentally observed.

B. Bundles of single walled nanotubes

As mentioned in the previous section, nanotubes are generally packed into bundles as presented in Figure 3. A schematic illustration of the structure is shown in Figure 12. The van der Waals interaction stabilizes the structure at a typical distance between the wall of the nanotubes about 3.14 Å.

C. Multi-walled nanotubes

Multi-walled nanotubes consist of at least two concentric nanotubes of different diameters like Matrioschka Russian doll. Figure 13 presents a 3 layers MWNT made of a $(6, 5)$ inside a $(10, 10)$ inside a $(16, 11)$. The geometrical properties of each individual layer can be described by the vector model.

1.5. Electronic Properties of Carbon Nanotubes

From the beginning, theoretical calculations have proposed that the electronic properties of the carbon nanotubes depend strongly on their geometrical structures.[3,25–27] Thus, nanotubes can be metals or semiconductors depending on their diameter or chirality. Figure 14 presents densities of states of different

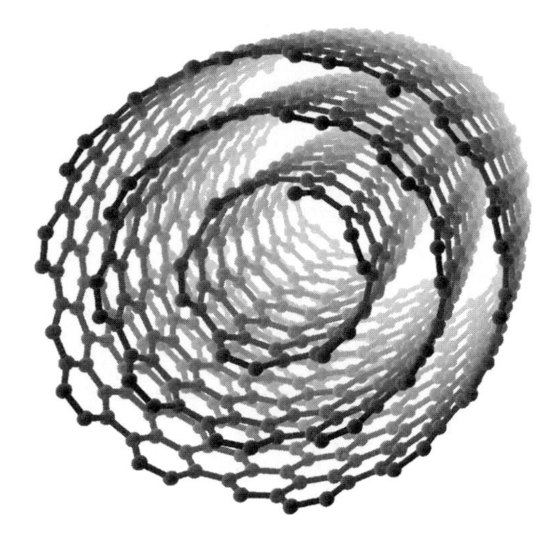

Fig. 13. A multi-walled carbon nanotubes : $(6,5)@(10,10)@(16,11)$.

type of nanotubes. Practically, from the pair of integer (n,m), it is possible to determine the electronic properties of the nanotubes. It was established, that if $n-m$ is a multiple of 3 the nanotube is metallic with a constant density of state over a large plateau above and below the Fermi level. In all others case, nanotubes are semiconductors with an energy gap inversely proportional to the radius R. Therefore, one third of the single walled carbon nanotubes are metallic with some exceptions for small diameters below 8 Å, because of their higher curvatures. This result is easy to remain and directly related to the periodic boundary conditions along the circumference of the nanotube which discretize the electronic states in one direction perpendicular to the nanotube axis.[28] Another interesting feature concerns the van Hove singularities related to the strong one dimensional character of the electronic states in the nanotubes. Actually, in the case of armchair nanotubes $m=n$ and small radius nanotubes detailed calculations and low temperature experiments[29] have show that the curvature opens a tiny gap. However, such pseudo gap can be ignored for properties and applications around room temperature.

1.6. Recent Works on Individual Carbon Nanotubes

The last developments of the research on nanotubes concern the studies of individual objects. From theory and modelization, an impressive number of

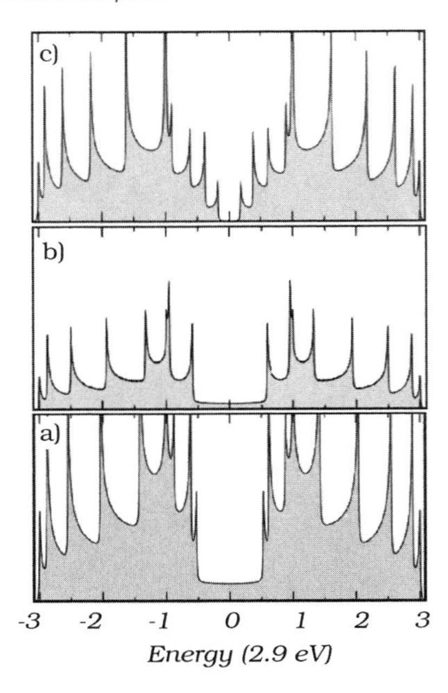

Fig. 14. Densities of states for two metallic carbon nanotubes (a) $(8, 2)$ and (b) $(5, 5)$. (c) is for the semiconducting $(10, 0)$. Adapted from Ref. 28.

remarkable properties have been predicted, suggesting a very broad potential of applications in energy storage, nano-electronics, mechanics, sensors, field emission.... However, it is a real challenge at the nanoscale to confront theory and experiments, since it is necessary to determine at the same time the structures, the mechanical, optical and electrical properties of one individual nanotube. This feat of strength has been initiated by the use of STM spectroscopy and high resolution transmission microscopy.[30,31] Figure 15 presents an atomically resolved STM measurement of a chiral carbon nanotube.[33] Later, quantum conductance behavior has been observed[30] along the axis of the tube and integration of carbon nanotubes as nano-junctions in logic circuit has been demonstrated.[32] A spatially controlled light emission from nanotube-field-effect transistor after the injection of electrons and holes from opposite edges was also reported.[34] The ultimate experiments have been performed on free-standing nanotubes as presented in Figure 16, where the two integer (n, m) have been determined by electron diffraction on several individual objects with diameter in the range of 14 to 30 Å and correlated to vibrational properties measured by micro-Raman.[35] If the knowledge on nanotubes is rapidly growing up, the field of their properties and applications is

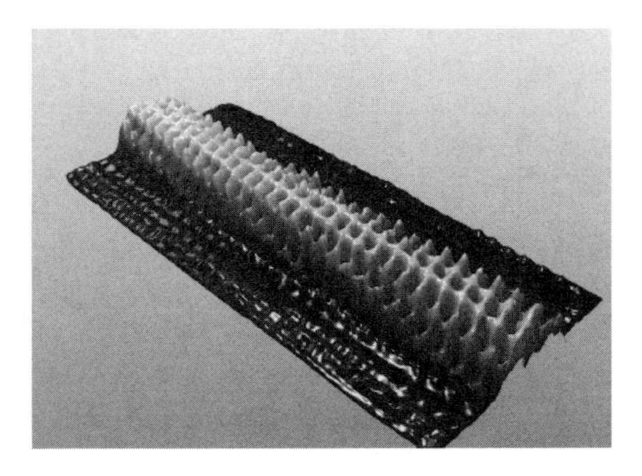

Fig. 15. Atomically resolved STM measurement of a chiral carbon nanotube.[33]

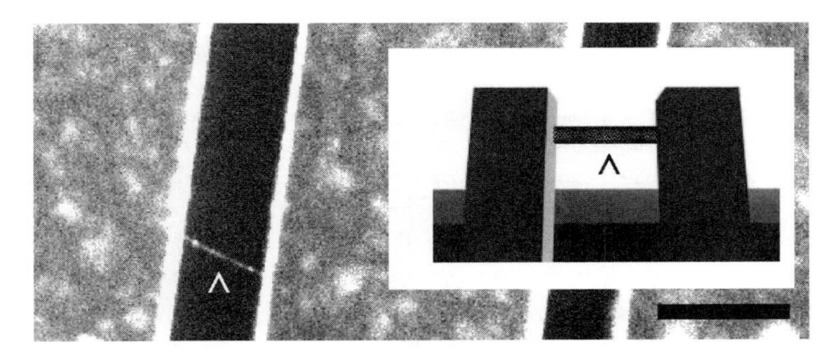

Fig. 16. Transmission Electron Micrograph of one suspended single walled carbon nanotube which has been investigated by electron diffraction and micro-Raman in order to determine its structure and vibrational properties.[35] The arrows indicate the free standing nanotube. The scale bar is 1 μm.

still largely unexplored. New nanostructures will be certainly discovered with their own characteristics as peapods[36] or chemically modified nanotubes by sidewall functionalization.[37]

Bibliography

1. Kroto, H. W., Heath, J. R., O'Brien, S. C., Curl, R. F. and Smalley, R. E. (1985) *Nature*, **318**, 162.
2. Krätschmer, W., Lamb, L. D., Fostiropoulos, K. and Huffman, D. R. (1990) *Nature*, **347**, 354.

3. Dresselhaus, M. S., Dresselhaus, G. and Eklund, P. C. (1996) *Science of Fullerenes and Carbon Nanotubes*, Academic Press, San Diego.
4. Kadish, K. M. and Ruoff, R. S. (2000). *Fullerenes: Chemistry, Physics and Technology*, Wiley-Interscience.
5. Iijima, S. (1991) *Nature*, **354**, 56.
6. Oberlin, A., Endo, M. and Koyama, T. (1976) *J. Cryst. Growth*, **32**, 335.
7. Millward, G. R. and Jefferson, D. A. (1978) *Chem. Phys. Carbon*, **14**, 1.
8. Iijima, S. (1980) *J. Microscopy*, **119**, 99.
9. Ebbesen, T. W. and Ajayan, P. M. (1992) *Nature*, **358**, 220.
10. Bethune, D. S., Kiang, C. H., de Vries, M. S., Groman, G., Savoy, R., Vasquez, J. and Beyers, R. (1993) *Nature*, **363**, 605.
11. Iijima, S. and Ichihashi, T. (1993) *Nature*, **363**, 603.
12. Thess, A., Lee, R., Nikolaev, P., Dai, H., Petit, P., Robert, J., Xu, C., Lee, Y. H., Kim, S. G., Rinzler, A. G., Colbert, D. T., Scuseria, G. E., Tomanek, D., Fischer, J. E. and Smalley, R. E. (1996) *Science*, **273**, 483.
13. Journet, C., Maser, W., Bernier, P., Loiseau, A., Lamy de la Chapelle, M., Lefrant, S., Deniard, P., Lee, R. and Fischer, J. E. (1997) *Nature*, **388**, 756.
14. Tersoff, J. and Ruoff, R. S. (1994) *Phys. Rev. Lett.*, **73**, 676.
15. Charlier, J. C., Gonze, X. and Michenaud, J. P. (1995) *Europhys. Lett.* **29**, 43.
16. Golberg, D., Bando, Y., Bourgeois, L. and Kurashima, K. (1999) *Carbon*, **39**, 1858.
17. Henrard, L., Loiseau, A., Journet, C. and Bernier, P. (2000) *Eur. Phys. J. B*, **13**, 661.
18. Guillard, T., Cetout, S., Alvarez, L., Sauvajol, J. L., Anglaret, E., Bernier, P., Flamant, G. and Laplaze, D. (1999) *Eur. Phys. J. B*, **5**, 252.
19. Endo, M., Takeuchi, K., Kobori, K., Takahashi, K., Kroto, H. W. and Sarkar, A. (1995) *Carbon*, **33**, 873.
20. Ivanov, V., Fonseca, A., Nagy, J. B., Lucas, A., Lambin, P., Bernaerts, D. and Zhang, X. B. (1995) *Carbon*, **33**, 1727.
21. Rinzler, A. G., Liu, J., Dai, H., Nikolaev, P., Huffman, C. B., Rodriguez-Macias, F. J., Boul, P. J., Lu, A. H., Heymann, D., Colbert, D. T., Lee, R. S., Fischer, J. E., Rao, A. M., Eklund, P. C., Smalley, R. E. (1998) *App. Phy. A*, **67**, 29.
22. Zhou, W., Ooi, Y. H., Russo, R., Papanek, P., Luzzi, D. E., Fischer, J. E., Bronikowski, M. J., Willis, P. A., Smalley, R. E. (2001) *Chem. Phys. Lett*, **350**, 6.
23. Kim, Y., Torrens, O., Kiikkawa, M., Abou-Hamad, E., Goze-Bac, C. and Luzzi, D. (2007) *Chem. Mater.*, **19**, 2982.
24. Goze-Bac, C., Latil, S., Lauginie, P., Jourdain, V., Conard, L., Duclaux, L., Rubio, A. and Bernier, P. (2002) *Carbon*, **40**, 1825.
25. Saito, R., Fujita, M., Dresslelhaus, G. and Dresselhaus, M. S. (1992) *Phys. Rev B*, **46**, 1804.
26. Dresselhaus, M. S., Dresselhaus, G. and Saito, R. (1995) *Carbon*, **33**, 883.
27. Hamada, N., Sawada, S. and Oshiyama, O. (1992) *Phys. Rev.Lett.*, **68**, 1579.
28. Charlier, J. C., Blase, X. and Roche, S. (2007) *Rev. Mod. Phys.*, **79**, 677.
29. Ouyang, M., Huang, J. L., Cheung, C. L. and Lieber, C. M. (2001) *Science*, **292**, 702.

30. Tans, S. J., Devoret, M. H., Dai, H., Thess, A., Smalley, R. E., Geerligs, L. J. and Dekker, C. (1997) *Nature*, **386**, 474.
31. Wildoer, J. W. G., Venema, L. C., Rinzler, A. G., Smalley, R. E. and Dekker, C. (1998) *Nature*, **391**, 59.
32. Collins, P. G., Arnold, M. S. and Avouris, P. (2001) *Science*, **292**, 706.
33. Lemay, S. G., Janssen, J. W., van den Hout, M., Mooij, M., Bronikowski, M. J., Willis, P. A., Smalley, R. E., Kouwenhovena, L. P. and Cees Dekker, (2001) *Nature*, **412**, 617.
34. Freitag, M., Chen, J., Tersoff, J., Tsang, J. C., Fu, Q., Liu, J. and Avouris, P. (2004) *Phys. Rev. Lett.*, **93**, 76803.
35. Meyer, C., Paillet, M., Michel, T., Moreac, A., Neuman, A., Duesberg, G., Roth, S., Sauvajol, J. (2005) *Phys. Rev. Lett.*, **95**, 217401.
36. Smith, B. W., Monthioux, M. and Luzzi, D. E. (1998) *Nature*, **396**, 323.
37. Hirsch, A. (2002) *Angew. Chem. Int.*, **41**, 1853.

2

Applications in Mechanics and Sensors

Angel Rubio

2.1. Introduction

Since the discovery of carbon nanotubes (CNTs) by Sumio Iijima[33] we witness an explosive development of the nanotube science and technology. Considering the rapid progress made in the fabrication, manipulation, characterization and modeling of nanostructures based on nanotubes, it is reasonable to expect that CNTs will pervade in the following years key application areas such as energy, materials, devices, and several others. There is a great variety of applications for which nanotubes represent a disruptive potential, ranging from energy storage, composites, nanoelectronics and other solid-state devices, to sensors and actuators (see for example the reviews by Baughman *et al.*[5]; Loiseau *et al.*[42]).

Several structural varieties of nanotubes have been identified and classified based on criteria such as helicity (also known as chirality), number of walls, inclusion of pentagons-heptagons, etc. The simplest form is the single-walled carbon nanotube (SWNT) which resembles a rolled honeycomb graphite layer into a mono-atomic-thick cylinder. Several concentrically embedded SWNTs form a multi-walled carbon nanotube (MWNT). Other nanotube varieties include nanotube bundles or ropes, inter-tube junctions, nanotori, coiled nanotubes, etc. Due to their relative simplicity and atomically precise morphology, single-walled carbon nanotubes offer the opportunity of assessing the validity of different macro- and microscopic models.

The properties of carbon nanotubes can be grouped into three categories: structural, mechanical and electronic. From the structural point of view, in most situations, CNTs can be considered one-dimensional (1D) objects, with typical diameters (dt) in the nm-range and lengths (L) reaching several micrometers. This one-dimensionality of tubes impacts on and is visible mostly through the mechanical and electronic properties. However, the structure of nanotubes can be exploited in itself such as for instance by field emitters or gas break-down sensors, which are based on the "sharpness" of CNTs giving rise to huge local electric fields.

While the prediction of electronic properties of carbon nanotubes required relatively subtle theoretical analysis, their unique mechanical behavior could be intuitively anticipated based on several features: strength of carbon bonds, their uniform arrangement within the graphitic sheet, and the seamless folding of this network into a tubule. The mechanical properties class is encompassing the elastic, thermal, vibrational or any other properties related to the motion of the tube's atoms. In nanotubes, carbon is sp^2-hybridized resulting in strong σ-bonds weakly reinforced by π-bonds. Considering the hybridization, it is natural to assume a certain overlap between nanotube and graphite (graphene) elastic properties, such as Young's modulus, bending, tensile and torsional stiffness, and yield strength. SWNTs have tensile moduli close to 1TPa (stiff as diamond) and strengths ≈50 GPa (corresponding to 5–10% maximal strain), which earned them the title of ultimate fibers. Despite their stiffness, CNTs retain a high bending flexibility due to their high aspect ratios (L /dt). With some exceptions, the thermal and vibrational properties of nanotubes also show similarities with graphite. Since the in-plane thermal conductivity of pyrolytic graphite is very high it is expected that the on-axis thermal conductivity of defect-free tubes would be even higher. At low temperatures the phonon mean free path is controlled mainly by boundary scattering and the coherence length ($\approx\mu$m) is larger in tubules than in high oriented pyrolytic graphite ($<0.1\ \mu$m).

The class of electronic properties practically encompasses all remaining properties. Therefore it contains transport and electric properties (classical-, spindependent- and super-conductivity, dielectric permittivity), optical properties (absorbtion, scattering, luminescence), magnetic properties (susceptibility, Zeeman splitting, Aharonov-Bohm effect), chemical properties (covalent and non-covalent binding), as well as hybrid properties and correlated manybody effects (thermopower, piezoresistivity, piezoelectricity, Coulomb blockade, Kondo effect, Tomonaga-Luttinger liquid behavior). The electronic properties of nanotubes are strongly modulated by small structural variations, in particular, their metallic or semiconducting character is determined by the diameter and helicity (chirality) of the carbon atoms in the tube. A recent review of electronic and transport properties of carbon nanotubes can be found in Ref. 12 and on the fundamentals and applications of nanotubes in the books by Loiseau *et al.*[42]; Ebbesen[24] and Dresselhaus *et al.*[22] (and references therein).

2.1.1. *Some Applications: Carbon Nanotube-Based Sensors*

Continuously expanding, the field of CNT-based sensors includes an already impressive list of demonstrators, encompassing (bio)chemical, strain, stress, pressure, mass, flow, thermal, and optical sensors. The underlying mechanism of sensors involves the modulation of a CNT's electronic properties.

In Refs. 38 and 18 it is proposed to use CNTs as sensitive materials for chemical sensors. In Ref. 38 the electrical conductance of a nanotube was found to increase when the tube was exposed to NO_2, and to decrease when exposed to NH_3. Electron charge transfer (doping) was proposed as the mechanism dictating the change in conductivity (DG) by shifting the Fermi level of the channel. Collins *et al.*[18] have shown that exposure to air or oxygen dramatically influences the nanotubes' electrical resistance, thermoelectric power, and local density of states, as determined by transport measurements and scanning tunneling spectroscopy. These electronic parameters can be reversibly "tuned" by surprisingly small concentrations of adsorbed gases, and semiconducting nanotubes can apparently be converted into metallic tubes through such exposure. Since these initial two reports, the list of chemFET-like gas and biological sensors has increased considerably. Innovations have also been brought, such as for example nanotube functionalization which has yielded highly sensitive and selective sensors.[50,60] Also, recently Goldsmith *et al.*[27] have demonstrated that it is possible to control the functionalization density by monitoring in real-time the conductance of a nanotube in an electrochemical setup.

Ghosh *et al.*[26] report flow sensors, in which the conductance of carbon nanotubes, disposed parallel to the flow lines, is clearly changed at different flow rates. The suggested mechanism for this nonlinear effect is forcing of the electrons in the nanotubes by the fluctuating Coulombic field of the liquid, or, briefly, Coulomb drag. Another CNT flow-meter demonstrated recently by Bourlon *et al.*[8] operates differently from the previous device. It is also worthy to mention that reading-out the modulation of the electronic properties in chemical sensor needs not be confined to electrical measurements. A promising approach is using light to probe either the photoabsorption or the fluorescence spectra.[11,29] Cao *et al.*[11] have studied band-gap photo-absorption in SWNTs, previously exposed to air, in different organic solvents.

An ionization CNT sensor has been demonstrated by Modi *et al.*[47] based on the ionization fingerprint which is characteristic to each analyte gas. This device simply exploits the geometry of the nanotubes, since their sharp tips generate very high electric fields at relatively low voltages, lowering gas breakdown voltages. Another category of sensor demonstrators that make use of CNTs as electrodes are the amperometric biosensors utilizing nanotubes as electrode material in electrochemical setups. This type of sensors are exhaustively reviewed in Wang,[67] Still other CNT sensing devices could be mentioned here, such as low temperature quantum electrometers[52] and superconducting quantum interference device (SQUID) magnetometers.[17]

2.2. Nanotube Nanomechanics

Carbon as well as composite BN nanotubes demonstrate very high stiffness to an axial load or a bending of small amplitude, which translates in the record-high efficient *linear*-elastic moduli. At larger strains, the nanotubes (especially, the single-walled type) are prone to buckling, kink forming and collapse, due to the hollow shell-like structure. These abrupt changes (bifurcations) manifest themselves as singularities in the *non-linear* stress-strain curve, but are *reversible* and involve no bond-breaking or atomic rearrangements. This resilience corresponds, quantitatively, to a very small sub-angstrom efficient thickness of the constituent graphitic shells. *Irreversible* yield of nanotubes begins at extremely high deformation (from several to dozens percent of in-plane strain, depending on the strain rate) and high temperature.[a] The atomic relaxation begins with the edge *dislocation dipole* nucleation, which (in case of carbon) involves a diatomic interchange, i. e. a 90° bond rotation.[b] A sequence of similar diatomic steps ultimately leads to failure of the nanotube filament. The failure threshold (yield strength) turns out to depend explicitly on nanotube *helicity*, which thus demonstrates again the profound role of symmetry for the physical properties, either electrical conductivity or mechanical strength. Finally, the manifestation of mechanical strength in the *multi-walled* or *bundled* nanotubes (ropes) is obscured by the poor load transfer from the exterior to the core of

[a]Note that the temperature is an important parameter in the strength of a material since the motion of dislocations is thermally activated. Like all covalent materials, nanotubes are brittle at low temperatures. The flexibility of the nanotubes at room temperature is due to their high strength and the unique ability of the hexagonal network to distort in order to release the applied stress.

[b]This bond-rotation defects are more unlikely in boron nitride due to the larger energy needed to form B–B or N–N bonds. This could increase the yield strength of the nanotube as compared to carbon.

such larger structure. This will lead to lower apparent strength and even lower linear moduli, as they become limited by the weak lateral interaction between the tubules rather than by their intrinsic carbon bond network. The ultimate strength of nanotubes and their ensembles is an issue that requires the modeling of inherently mesoscopic phenomena, such as plasticity and fracture, on a microscopic, atomistic level, and constitutes a challenge from the theoretical as well as experimental points of view. A recent very nice overview assessing carbon nanotube strength is given in Dumitrica *et al.*[23]

The similarities among graphite and other sp^2-like bonded materials as hexagonal boron nitride and boron-carbon-nitrogen compounds, lead us to theoretical proposition that $B_xC_yN_z$ nanotubes would be stable.[43,44,53,54] Specific synthesis of these nanotubes was achieved afterwards: boron-nitride[15,41] and BC_2N and BC_3[61,68] as well as other inorganic tubular forms of WS_2 and MoS_2.[62] The predicted properties of these tubules are quite different from those of carbon with numerous possible technological applications in the fields of catalysis, lubrication, electronic and photonic devices.[43,44,c]

Calculations of the stiffness of SWNTs demonstrated that the Young modulus shows a small dependence on the tube diameter and chirality for the experimental range of nanotube diameters (between 1.3 and 1.4 nm). It is predicted that carbon nanotubes have the highest Young's modulus of all the different types of composite tubes considered (BN, BC_3, BC_2N, C_3N_4, CN).[31] Those results for the C and BN nanotubes are reproduced in the left panel of Fig. 1. Furthermore, the Young modulus approaches, from below, the graphitic limit for diameters of the order of 1.2 nm. The computed value of C for the wider carbon nanotubes of 0.43 TPa nm, that corresponds to 1.26 TPa modulus in our convention, is in excellent agreement with the experimental value for SWNT's of 1.25 TPa.[66] It is also in rather good agreement with the value of 1.28 TPa reported for multi-wall nanotubes (MWNT).[67] Although this result is for MWNT the similarity between SWNT is not surprising as the intra-wall C–C bonds mainly determine the Young's modulus. From these results we can

---

[c]The electronic properties of BN nanotubes are quite different to carbon, namely: all are stable wide band-gap semiconductors independent of helicity and diameter of the nanotube and of whether the nanotube is single- or multi-walled. On the other hand, single-wall BC_3 tubules are found to be semiconductor with a small gap of $\sim$0.5 eV that would disappear in the multilayer form.[43,44] The inter-wall interaction for concentric tubules makes the conduction band overlap with the valence band maximum (σ-bands). Therefore, a concentric needle of BC_3 tubules will have σ conductivity and opens the question of how many walls are necessary to give rise to conductivity in concentric BC_3 tubules, that is to look for a semiconductor-metal transition as a function of the number of concentric shells.

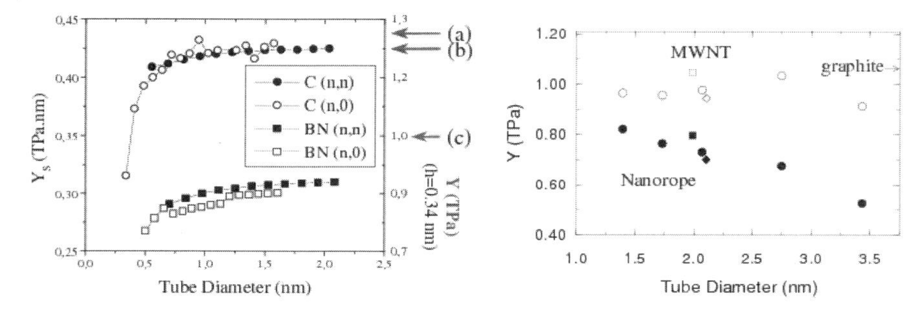

Fig. 1. Left: Young modulus for armchair and zig-zag carbon and BN nanotubes. The values are given in the proper unit of TPa nm for SWNTs (left axis), and converted to TPa (right axis) by taking a value for the graphene thickness of 0.34 nm. The experimental values for carbon nanotubes are reported on the right-hand-side: (a) 1.28 TPa[69]; (b) 1.25 TPa[39]; (c) 1 TPa for MWNTs.[48] Right: Young modulus versus tube diameter in different arrangements. Open symbols correspond to the multi-wall geometry (10-layer tube with an interlayer distance of 0.34 nm), and solid symbols for the single-wall-nanotube crystalline-rope configuration. In the MWNT geometry the value of the Young modulus does not depend on the specific number of layers. The experimental value of the $c_{11} = 1.06$ TPa elastic constant of graphite is also shown.[58]

estimate the Young modulus considering two different geometries of practical relevance:

(1) *multi-wall* like geometry, in which the normal area is calculated using the wall-wall distance as the one in MWNTs, which is very approximately equal to the one of graphite,
(2) *nanorope or bundle* configuration of SWNTs, where the tubes would be arranged forming an hexagonal closed packed lattice, with a lattice constant of $(2R + 3.4)$, being R the tube radius.

The results for these two cases are presented in the right panel of Fig. 1. The MWNT geometry give a value that is very close to the graphitic one, however the rope geometry shows a decrease of the Young modulus with increasing the tube radius due to the quadratic increase of the effective area in this configuration, while the number of atoms increases only linearly with the tube diameter. The computed values for the SWNT ropes experimentally observed are, however, still very high (0.5 TPa), even comparing with other known carbon fibers. This value is in quite good agreement with AFM experiments on anchored SWNTs ropes[57] ($Y \sim 0.6$ TPa) and for stress-strain puller measurements of the Young modulus for aligned nanotube ropes of MWNTs ($Y \sim 0.45 \pm 0.23$ TPa).[49] We should remark that this apparent "low" Y is not due to the presence of defects as arguments in Pan,[49] and can be understood in simple geometrical terms by the particular "empty" cross-sectional area is responsible for the apparent lower modulus.

For composite nanotubes, the results of Chopra[16] for the Young modulus of BN MWNT's of 1.22 TPa, which is somewhat larger than the result obtained for these tubes in the TB calculations (~0.9 TPa),[30] but nevertheless the agreement is close.[d] We indicate that BN and BC_3 tubes have similar values of the Young modulus,[e] although the latter have slightly larger values. In those studies C_3N_4 nanotubes are shown to be much softer than any other type of tube, the reason being that for a given amount of tube surface, there is a smaller density of chemical bonds.[31]

The Poisson ratio is given by the variation of the radius of the SWNT resulting from longitudinal deformations along the tube axis. In all cases the computed Poisson ratio is positive: an elongation of the tube reduces its diameter. *The ab initio* values are 0.14 (from 0.12 to 0.16) for the armchair (n,n) tubes, and a little larger for other chiralities: 0.19 for (10,0) and 0.18 for (8,4). The uncertainty of the obtained values is of the order of 10%. In summary, the *ab initio* calculations indicate that the Poisson ratio retains graphitic values except for a possible slight reduction for small radii. It shows chirality dependence: (n,n) tubes display smaller values than (10,0) and (8,4). Similar differences are found between the *ab initio* and TB calculations for BN tubes, namely for the (6,6) tube the *ab initio* value for the Poisson ratio is 0.23 whereas the TB one is 0.30.[30]

Large amplitude deformations, beyond the Hookean behavior, reveal nonlinear properties of nanotubes, unusual for other molecules or for the graphite fibers. Both theory-simulations and experimental evidence suggest the ability of nanotubes to significantly change their shape, accommodating to external forces without irreversible atomic rearrangements. They develop kinks or the ripples (multi-walled tubes) in compression and bending, flatten into deflated "ribbons" under torsion, and still can reversibly restore original shape. This *resilience* is unexpected for a graphite-like material. It must be attributed to the small dimension of the tubules, which leaves no room for the stress-concentrators — micro-cracks or dislocation piles, making a macroscopic material prone to failure. Furthermore, MWNT can be bent repeatedly through large angles using the tip of an atomic force microscope without undergoing catastrophic failure.[25] The observed response at very high strain deformation

---

[d]The smaller Young modulus for BN compared to C-nanotubes is directly related to the difference in the experimental c_{11} elastic constants that are 0.75 TPa and 1.06 TPa for hexagonal-BN and graphite, respectively.

[e]The calculated average Young modulus is 0.9 TPa and 0.92 Tpa for BN and BC_3, respectively.

indicates that nanotubes are remarkably flexible and resilient. The main outcome of mechanical studies of carbon nanotubes is that both thin- and thick-walled carbon nanotubes exhibit compressive strengths about two orders of magnitude higher than any other known fiber.[f]

2.2.1. *Experimental Evidence of Nanotube Resilience*

Collapsed forms of the nanotube ("nanoribbons") have been observed in experiment (Fig. 2d) and their stability can be explained by the competition between the van der Waals attraction and elastic energy. The basic physics can be understood by noticing that the elastic curvature energy per unit length is proportional to 1/R (R, radii of the tube); however, for a fully collapsed single-wall tubule with the opposite tubule walls at the typical van der Waals contact distance c, the energy per unit length would be composed of a higher curvature energy due to the edges which is independent of the initial tubule diameter,

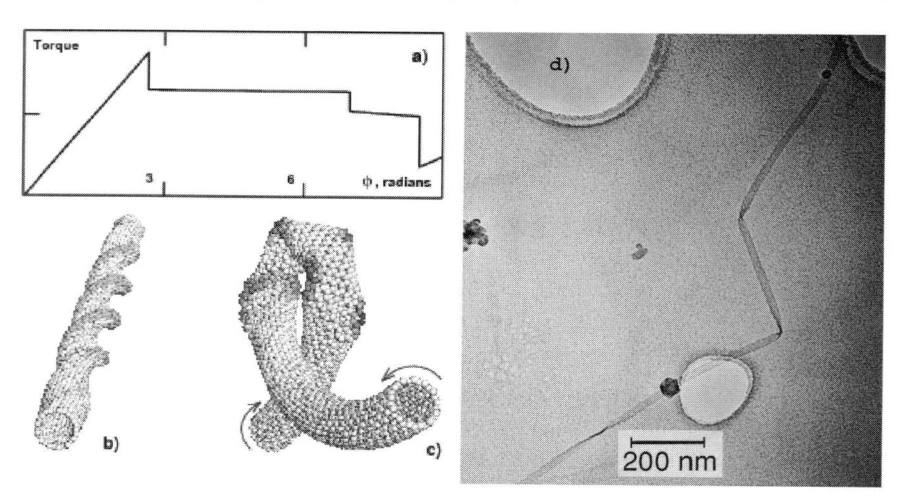

Fig. 2. Simulation of torsion and collapse.[71] The strain energy of a 25 nm long (13,0) tube as a function of torsion angle ϕ (a). At first bifurcation the cylinder flattens into a straight spiral (b) and then the entire helix buckles sideways, and coils in a forced tertiary structure (c). Collapsed tube (d) as observed in experiment.[14,15]

---

[f] Calculations of the elastic properties of C-nanotubes confirm that they are extremely rigid in the axial direction (high-tensile strength) and more readily distort in the perpendicular direction (radial deformations), due to their high aspect ratio. The detailed studies, stimulated first by experimental reports of visible kinks, lead one to conclude that, in spite of their molecular size, nanotubes obey very well the laws of continuum shell theory.[3,40,65]

and a negative van der Waals contribution, ε_{vdW} 0.03-0.04 eV/atom, that is $\propto$ R per unit length. Collapse occurs when the latter term prevails above a certain critical tube radii R_c that depends on the number N of shells of the nanotube, $R_c(N = 1) \sim 8c$ and $R_c(N = 8) \sim 19c$,[14] and the thickness of the collapsed strip-ribbon is (2N-1)c. Any additional torsional strain imposed on a tube in experimental environment also favors flattening[70–73] and facilitates the collapse, (Figs. 2b–c).

The bending seems fully reversible up to very large bending angles despite the occurrence of kinks and highly strained tubule regions in simulations, that are in excellent morphological agreement with the experimental images.[33] Similar bent-buckled shapes have been reported by several groups[20,33,55] (Fig. 2). This apparent flexibility stems from the ability of the sp^2 network to rehybridize when deformed out of plane, the degree of sp^2-sp^3 rehybridization being proportional to the local curvature.[28]

The accumulated evidence thus suggests that the strength of the carbon-carbon bond does not guarantee resistance to radial, normal to the graphene plane deformations. In fact, the graphitic sheets of the nanotubes, or of a plane graphite[32] though difficult to stretch are easy to bend and to deform. A crystal-array[64] or a pair[56] of parallel nanotubes flatten at the lines of contact between them so as to maximize the attractive van der Waals intertube interaction. *Computer simulations* have provided a compelling evidence of the mechanisms discussed above. By carefully tuning the tension in the tubule and gradually elevating its temperature, with extensive periods of MD annealing, the first stages of the mechanical yield of CNT have been observed.[22] In simulation of tensile load the novel patterns in plasticity and breakage, just described above, has clearly emerged.

2.3. Applications of Mechanical Response

The understanding of the mechanical response of nanotubes to external forces is of relevance for the application of nanotubes as a composite material reinforcement as well as in electronic devices, where the deformation of the tubes induced by the substrate alter locally the electronic properties of the nanotube. A broad discussion of potential applications of the nanotube can be found in existing reviews,[1,6,24,73] also outlining the challenges of implementation. There are two already *actual* applications, where the carbon nanotube can *commercially* compete with other materials units, and the mechanics of carbon nanotube plays either central or an important secondary role in both cases.

Different to traditional graphitic fibers, nanotubes combine high flexibility and high strength with high stiffness. These properties open the way for a new generation of high performance light-weight polymer composites composites useful for structural reinforcement. This has triggered a lot of research looking at the ability of nanotubes to stiffen and strengthen a polymer.[1,59] The reinforcement will depend on how load is transferred to the nanotube aggregates such as single-walled nanotube bundles. If the adhesion between the matrix and the nanotubes is not strong enough to sustain high loads, the benefit of the high tensile strength of carbon nanotubes are lost. Load transfer in carbon nanotube epoxy composites was studied in both tension and compression.[59] Under tension it seems that only the outermost nanotube is loaded due to weak interlayer bonding. This could be related to the absence of registry between graphene layers that makes adhesion and friction decresed.[25] The excepcional mechanical properties of nanotubes will be reflected in composites once a good load transfer between the matrices and outer surface of the nanotubes is obtained.

Nanotubes are ideal proximal tips because the do not plastically deform during tip crashes on the surfaces as conventional tips often do. Instead they elastically deform, buckle and slip. The recovering mechanism has to be taken into account when analyzing the data from AFM measurements using the nanotubes as tips. The enhanced capability of a scanning force microscope using carbon nanotube has been already demonstrated.[19] Nanotube tips have also been used as pencils for writing 10-nm-width structures on silicon substrates. The robustness of the nanotube tips permits a writing rate 0.5 mm/sec, five times faster than was possible with older AFM tips. The way the nanotube writes is for an electric field, emanating from the nanotube, to remove hydrogen atoms from a layer of hydrogen atop a silicon base. The exposed silicon surface oxidizes; thus the "writing" consists of narrow SiO_2 track.

The development of new tools to manipulate and analyze the nanoworld relies in having traditional macrocopic techniques down to the nanoscale. Just recently it has been probed that the mechanical robustness and electrical conductivity of nanotubes can be used to build electromechanical-tweezers.[37] The nanotubes are attached to a metal electrode. This tweezers not only allow to grab and manipulate mesoscopic silicon-carbide dots and gallium-arsenide wires but also can be used to probe electronic properties across the structure (transport). Many other applications can be envisioned, for example, the manipulation of biological structures.

A full-sealed field emission display has been recently reported by the group at Samsung Corporation[13] who used a relatively disordered array of

SWNT embedded in a composite substrate, in combination with closely-placed (200 mm) phosphorous layer. At the SWNT density of only 1-3 mm^2, the current density is stable for hours and high enough to induce bright light. In this important example, mechanical robustness of the carbon network in the tips prevents them from deterioration.

Mechanical behavior is often coupled with other physico-chemical phenomena, which of course broadens the import of mechanical properties, for example Elasto-electronics: Coupling with electrical properties can manifest itself both ways, where either deformation affects the charge distribution or transport, or an additional charge can cause visible deformation, thus making an actuator. Change in the bandgap of nanotubes with strain and torsion has been discussed.[36,g] This variation could have some implications for nanoscale electro-mechanical devices.[35]

Furthermore, bending of a carbon nanotube introduces an increased mixing of σ and π states that leads to an enhanced density of states at the Fermi energy region and to a charge polarization of the carbon-carbon bond in the deformed region.[51] The transport properties of bent tubes depend on the chirality of the tube, indeed, armchair tubes keep their metallic character irrespective of the deformation but this is not the case for chiral metallic tubes where local-electronic barriers arise from the bending of the tube.[9,10] This calculations are in agreement with recent experimental transport data of individual carbon nanotubes supported on a series of electrodes.[7]

Mechanical stretching has been proposed as a method for inducing chiral conductivity in these carbon nanotubes when doped.[45] The essence of chiral conduction is symmetry breaking on tubule walls upon mechanical stretching. Calculations of the transport properties of mechanically stretched doped-carbon nanotubes show that stretching induces chirality on the tubule current, being the most efficient geometry for the induced chirality an arm-chair like atomic arrangement. If a nanocoil is subjected to an external frequency-dependent electric field the effective inductance varies with respect to AC-frequency different to the inductances of ordinary coils. This induced current chirality is most likely for geometrically chiral nanotube having helical pitches close to those of armchair tubules. By solving Maxwell's equations for chiral conducting nanotubes (nanocoils) it is found that the self-inductance and the resistivity of nanocoils should depend on the frequency of the alternating

---

[g]The coupling between the conduction electrons and long wavelength twistons, i.e., torsional shape vibrations is relevant for understanding the transport properties at low temperatures, namely, the relevance for the peculiar linear dependence of the electrical resistivity with temperature in the metallic armchair tubes.

current even when the capacitance of the nanocoils is not taken into account.[46] This is in contrast to elementary treatment of ordinary coils. This fact is useful to distinguish nanocoils by electrical measurements.[h]

An interesting manifestation of electro-mechanical coupling in nanotubes has recently emerged as carbon nanotube actuators.[4] By changing the applied voltage and therefore injecting the electrons or holes one causes either an expansion or contraction of carbon nanotube, or a graphene sheet. A separate contribution of a purely quantum-mechanical change in band structure and in orbital occupancy and the role of electrochemical double-layer are not completely understood, the experimental evidence is quite convincing. Nanotube sheets adhered to the opposite sides of insulating film make a cantilever sensitive to the voltage applied between the sheets: a small elongation on one side and a contraction on the other results in significant bending of the bilayer cantilever. The limit of the gravimetric work capacity, $1/2\ \mathbf{Y}\varepsilon^2/\rho$, is expected to be much higher than for other materials, like ferroelectrics, due to great values of modulus $\mathbf{Y} = 1$ TPa, strain range $\varepsilon = 1\%$, and the low density $\rho = 1.3\ \text{g/cm}^3$ for carbon nanotube bundles.

Acknowledgements

We are grateful to S. Roche, C. Roman, V. Crespi, M. Cohen, S. G. Louie, P. M. Ajayan and acknowledge funding from the European Community through NoE Nanoquanta (NMP4-CT-2004-500198), SANES (NMP4-CT-2006-017310), DNA-NANODEVICES (IST-2006-029192) and NANO-ERAChemistry projects, UPV/EHU (SGIker Arina) and Basque Country University (SGIKer ARINA) and "Grupos Consolidados UPV/EHU del Gobierno Vasco" (IT-319-07) and e-I3 ETSF project (INFRA-2007-1.2.2: Grant Agreement Number 211956). We acknowledges the computer resources, technical expertise and assistance provided by the Barcelona Supercomputing Center-Centro Nacional de Supercomputación.

Bibliography

1. Ajayan, P. M. and Ebbesen, T. W. (1997) *Reports on Progress in Physics*, **60**, 1025.
2. Ajayan, P. M., Schadler, L. S., Giannaris, C. and Rubio, A. (2000) *Adv. Mat.*, **12**, 750.

---

[h] The basic physics behind this phenomenon is that in a nanocoil the chiral angle of the current is frequency dependent whereas in a classical coil this is fixed by the pitch of the winding of the wires in forming the coil.

3. Allen, H. G. and Bulson, P. S. (1980) *Background to Buckling*, McGraw-Hill, London, p. 582.
4. Baugman, R. H., *et al.* (1999) *Science*, **284**, 1340.
5. Baugman, R. H., *et al.* (2002) *Science*, **297**, 787.
6. Bernholc, J., Roland, C. and Yakobson, B. I. (1997) *Current Opinion in Solid State & Materials Science*, **2**, 706–715.
7. Bezryadin, A., Verschueren, A. R. M., Tans, S. J. and Dekker, C. (1999) *Phys. Rev. Lett.*, **80**, 4036.
8. Bourlon, B., Wong, J., Mikó, C., Forró, L. and Bockrath, M. (2007) *Nature Nanotech.*, **2**, 104.
9. Buongiorno-Nardelli, M., Yakobson, B. I. and Bernhold, J. (1998) *Phys. Rev. B*, **57**, 4277.
10. Buongiorno-Nardelli, M., Yakobson, B. I. and Bernhold, J. (1998) *Phys. Rev. Lett.*, **81**, 4656.
11. Cao, A., Talapatra, S., Choi, Y. Y., Vajtai, R., Ajayan, P. M., Filin, A., Persans, P. and Rubio, A. (2005) *Advanced Materials*, **17**, 147.
12. Charlier, J.-C., Blasé, X. and Roche, S. (2007) *Rev. Mod. Phys.*, **79**, 677.
13. Choi, W. B., *et al.* (1999) *Appl. Phys. Lett.*, **75**, 3129.
14. Chopra, N. G., Benedict, L. X., Crespi, V. H., Cohen, M. L., Louie, S. G. and Zett, A. (1995a) *Nature*, **377**, 135.
15. Chopra, N. G., Luyken, R. J., Cherrey, K., Crespi, V. H., Cohen, M. L., Louie, S. G. and Zettl, A. (1995b) *Science*, **269**, 966.
16. Chopra, N. G. and Zettl, A. (1998) *Solid State Comm.*, **105**, 297.
17. Cleuziou, J.-P., Wernsdorfer, W., Bouchiat, V., Ondarçuhu, T. and Monthioux, M. (2006) *Natl. Nanotech.*, **1**, 53.
18. Collins, P. G., Bradley, K., Ishigami, M. and Zettl, A. (2000) *Science*, **287**, 1801.
19. Dai, H., Hafner, J. H., Rinzler, A. G., Colbert, D. T. and Smalley, R. E. (1996) *Nature*, **384**, 147.
20. Despres, J. F., Daguerre, E. and Lafdi, K. (1995) *Carbon*, **33**, 87.
21. Dresselhaus, M. S., Dresselhaus, G. and Eklund, P. C. (1996) *Science of Fullerenes and Carbon Nanotubes*, Academic Press, San Diego, p. 965.
22. Dresselhaus, M. S., Dresselhaus, G., Sugihara, K., Spain, I. L. and Goldberg, H. A. (1998) *Graphite Fibers and Filaments*, Springer-Verlag, Heidelberg, p. 382.
23. Dumitrica, T., Hua, M. and Yakobson, B. (2006) *Proc. Natl. Acad. Sci.*, **103**, 6105.
24. Ebbesen, T. W., ed., (1997) *Carbon Nanotubes: Preparation and Properties*, CRC Press, Tokyo, p. 296.
25. Falvo, M. R., Clary, G. J., Taylor, R. M., Chi, V., Brooks, F. P., Washburn, S. and Superfine, R. (1997) *Nature*, **389**, 582.
26. Ghosh, S., Sood, A. K. and Kumar, N. (2003) *Science*, **299**, 1042.
27. Goldsmith, B. R., *et al.* (2007) *Science*, **315**, 77.
28. Haddon, R. C. (1993) *Science*, **261**, 1545.
29. Heller, D. A., *et al.* (2006) *Science*, **311**, 508.

30. Hernández, E., Goze, C., Bernier, P. and Rubio, A. (1998) *Phys. Rev. Lett.*, **80**, 4502.
31. Hernández, E., Goze, C., Bernier, P. and Rubio, A. (1999) *Appl. Phys. A.*, **68**, 287.
32. Hiura, H., Ebbesen, T. W., Fujita, J., Tanigaki, K. and Takada, T. (1994) *Nature*, **367**, 148.
33. Iijima, S. (1991) *Nature*, **354**, 56.
34. Iijima, S., Brabec, C., Maiti, A. and Bernholc, J. (1996) *J. Chem. Phys.*, **104**, 2089.
35. Joachim, C. and Gimzewski, J. K. (1997) *Chem. Phys. Lett.*, **265**, 353.
36. Kane, C. L. and Mele, E. J. (1997) *Phys. Rev. Lett.*, **78**, 1932.
37. Kim, P. and Lieber, C. M. (1999) *Science*, **286**, 2149.
38. Kong, J., *et al.* (2000) *Science*, **287**, 622.
39. Krishnan, A., Dujardin, E., Ebbesen, T. W., Yanilos, P. N. and Treacy, M. M. J. (1998) *Phys. Rev. B*, **58**, 14013.
40. Landau, L. D. and Lifschitz, E. M. (1975) *Theory of Elasticity*, 3rd ed., Pergamon, Oxford.
41. Loiseau, A., Willaime, F., Demoncy, N., Hug, G. and Pascard, H. (1996) *Phys. Rev. Lett.*, **76**, 4737.
42. Loiseau, A., Launois, P., Pteit, P., Roche, S. and Salvetat, J.-P. (2006) Understanding Carbon Nanotubes, *Lecture Notes in Physics*, Springer Verlag, Vol. 677.
43. Miyamoto, Y., Rubio, A., Louie, S. G. and Cohen, M. L. (1994a). *Phys. Rev. B*, **50**, 4976.
44. Miyamoto, Y., Rubio, A., Louie, S. G. and Cohen, M. L. (1994b) *Phys. Rev. B*, **50** 18360.
45. Miyamoto, Y. (1996) *Phys. Rev. B*, **54**, R11149.
46. Miyamoto, Y., Rubio, A., Louie, S. G. and Cohen, M. L. (1999) *Phys. Rev. B*, **60**, 13885.
47. Modi, A., Koratkar, N., Lass, E., Wei, B. and Ajayan, P. M. (2003) *Nature*, **424**, 171.
48. Muster, J., Burghard, M., Roth, S., Düsberg, G. S., Hernández, E. and Rubio, A. (1998) *J. Vac. Sci. Technol.*, **16**, 2796.
49. Pan, Z. W., Xie, S. S., Lu, L., Chang, B. H., Sun, L. F., Zhou, W. Y., Wang, G. and Zhang, D. L. (1999) *Appl. Phys. Lett.*, **74**, 3152.
50. Qi, P., *et al.* (2003) *Nano Lett.*, **3**, 347.
51. Roschefort, A., Avouris, P. and Salahub, D. R. (1999) *Phys. Rev. B*, **60**, 13824.
52. Roschier, L., Tarkiainen, R., Ahlskog, M., Paalanen, M. and Hakonen, P. (2001) *Appl. Phys. Lett.*, **78**, 3295.
53. Rubio, A., Corkill, J. L. and Cohen, M. L. (1994) *Phys. Rev. B*, **49**, 5081.
54. Rubio, A. (1997) *Cond. Matt. News*, **6**, 6.
55. Ruoff, R. S. and Lorents, D. C. (1995) *Bulletin of the APS*, **40**, 173.
56. Ruoff, R. S., Tersoff, J., Lorents, D. C., Subramoney, S. and Chan, B. (1993) *Nature*, **364**, 514.
57. Salvetat, J. P., Briggs, G. A. D., Bonard, J. M., Bacsa, R. R., Kulik, A. J., Stöckli, T., Burnham, N. A. and Forro, L. (1999) *Phys. Rev. Lett.*, **82**, 944.
58. Sánchez-Portal, D., Artacho, E., Soler, J. M., Rubio, A. and Ordejón, P. (1999) *Phys. Rev. B*, **59**, 12678.

59. Schadler, L. S., Giannaris, S. C. and Ajayan, P. M. (1998) *Appl. Phys. Lett.*
60. Staii, C. and Johnson, A. T. (2005) *Nano Lett.*, **5**, 1774.
61. Stephan, O., Ajayan, P. M., Colliex, C., Redlich, P., Lambert, J. M., Bernier, P. and Lefin, P. (1994) *Science*, **266**, 1683.
62. Tenne, R., Margulis, L., Genut, M. and Hodes, G. (1992) *Nature*, **360**, 444.
63. Tenne, R. (1995) *Adv. Mater.*, **7**, 965; and references therein.
64. Tersoff, J. and Ruoff, R. S. (1994) *Phys. Rev. Lett.*, **73**, 676.
65. Timoshenko, S. P. and Gere, J. M. (1998) *Theory of Elastic Stability*, McGraw-Hill, New York, p. 541.
66. Treacy, M. M. J., Ebbesen, T. W. and Gibson, J. M. (1996) *Nature*, **381**, 678.
67. Wang, J. (2005) *Electroanalysis*, **17**, 7.
68. Weng-Sieh, Z., Cherrey, K., Chopra, N. G., Blase, X., Miyamoto, Y., Rubio, A., Cohen, M. L., Louie, S. G., Zettl, A. and Gronsky, R. (1995) *Phys. Rev. B*, **51**, 11229.
69. Wong, E. W., Sheehan, P. E. and Lieber, C. M. (1997) *Science*, **277**, 1971.
70. Yakobson, B. I., Brabec, C. J. and Bernholc, J. (1996a) *Phys. Rev. Lett.*, **76**, 2511.
71. Yakobson, B. I., Brabec, C. J. and Bernholc, J. (1996b) *J. Computer-Aided Materials Design*, **3**, 173.
72. Yakobson, B. I., Campbell, M. P., Brabec, C. J. and Bernholc, J. (1997a) *Computational Materials Science*, **8**, 341.
73. Yakobson, B. I. and Smalley, R. E. (1997b) *American Scientist*, **85**, 324.

3
Applications in Biology and Medicine

Guido Rasi and Claudia Matteucci

Nanotechnology has become a popular term representing the main efforts of the current science and technology. The applications of this new technology seem to be finding its way in the products of almost every sector of manufacturing and consumer goods. Nanotechnology is also a bottom-up approach that focuses on assembling simple elements to form complex structures. The molecular structure at the nanoscale could provide properties extremely different from the same chemical substance in bulk form.

One of the important areas of nanotechnology is nanomedicine, defined as the application of nanotechnology to health. The European Technology Platform on Nanomedicine has highlighted the potentials of nanomedicine for revolutionizing the medical intervention at the molecular scale for diagnosis, prevention and treatment of diseases.[10] The impact is expected to be relevant not only in the social and welfare aspect but also for the economic as well. In fact, the ageing population and, the high expectations for better quality of life and the changing lifestyle call for continuous improvement in health care managment. Nanotechnology develops physical, chemical, and biological properties of materials at the nanometric scale.

Carbon nanotubes (CNTs) are one of the more representative materials for nanotech application. In their simplest form CNTs consist of one rolled sheet, known as single-walled carbon nanotubes (SWNT) with diameter ranging from around 1–10 nm, as well as concentric cylinders of carbon atoms known as multi-walled carbon nanotubes (MWNT) with diameter ranging from around 5–30 and length from nanometers up to millimetres. This last form of nanotubes supply an exceptionally high surface area to volume ratio, and at the same time, upon different orientation of the cylinders they have electrical properties ranging from those of a metal to those of semiconductor. Hence, CNTs possess a very broad range of electronic, thermal, and structural properties defined by diameter, length, and chirality or twist.[5,25,35,46] This versatility has given rise to a wide variety of applications, including many in the biological and medical field.

Owing to their nanometer dimensions, CNTs have the potential to interact at the cellular and molecular level. CNTs are in fact available for various chemical functionalisation strategies that improve their solubility and biocompatibility, allowing CNTs to represent new and promising composites in a wide variety of biomedical applications, as, cellular imaging, chemical and biological sensing, drug delivery, and matrix engineering for regenerative medicine. This chapter reviews the applications of CNTs technology in biology and medicine, providing examples and comments of the most promising application.

Applications in both, diagnosis and therapy are considered. As expected, the border between the two is difficult to design and it will be more and more undefined, as long as the molecular medicine will take place. The "theragnostic" is already a growing field.

FDA (U.S. Food and Drug Administration) funded a program for the co-development of drugs and the related biomarkers, also based on the experience that the molecular target may be the terminal of both the diagnostic approach (molecular imaging) and the therapeutic tool. The new targeted therapy using monoclonal antibodies against HER-2 protein, in breast cancer, and against EGFR (epidermal growth factor receptor) in colorectal cancer, represent two clear examples of the future scenery. In this context CNTs are powerful candidate to play a pivotal role and the examples here provided should be considered in this view. The following paragraph addresses "Biosensor Devices" together with imaging and tissue engineering, and gives body to the expectation about the potentiality and versatility described for this technology. The successive paragraph addresses therapy, reporting about "Drug, Peptide and Gene Delivery". Additionally will be presented an overview of the possibilities to exploit the CNTs technology in the fight against cancer, by assembling or combining the different applications previously described and allowing for multitasking effect.

3.1. Application in Biosensing, Imaging and Tissue Engineering

Since the discovery of fullerenes by Smalley, carbon nanotubes, and synthesis of crystalline semiconductor nanowires, innovative research for applications within molecular electronics and chemical and biological sensing has intensified.[13,19] Studies on developing highly sensitive biosensors, aiming at the early detection and clinical diagnoses of various diseases, are at present widely studied. Despite the high-quality of traditional techniques for proteins and biological molecules detection, the needs to continuously improve the rapidity of tests and to have real time results are highly warranted. Owing to their superior electronic and mechanical properties along with nanoscale dimensions, CNTs offer the basic components to develop extremely sensitive biosensors, that could provide real time data related to the physiological relevant parameters such as pH, pO_2, and glucose levels. Thus, nanosensors could be used to monitocellular physiology and behaviour such as ion transport, enzyme interactions, protein and metabolite secretions, with advantages in disease diagnosis and therapy response. Nanosensors could also be used in tissue engineering.[12,33]

CNT-based sensors have in fact already been demonstrated to measure many biological relevant factors. Amperometric glucose biosensors based on platinum nanoparticle-modified carbon nanotube electrode were investigated. The nanocomposite electrode exhibited high sensitivity, high selectivity, and low detection limit in amperometric glucose sensing at physiological neutral pH.[8,40,47] MWNTs have already been shown to be able to monitor insulin levels with the possibility to evaluate pancreatic islet cells before implantation into a diabetic patient.[44,51] Other authors have studied CNT-modified biosensor for monitoring total cholesterol in blood.[21] The possibility of placing the biosensor onto a biocompatible substrate is also appealing for detection in biological fluids. Tan X. *et al.* were able to measure free cholesterol in blood using a multiwalled carbon nanotube electrode placed on a biocompatible substrate.[39]

Zhang F.-F. *et al.* have described a novel reagent-less amperometric uric acid biosensor based on functionalized MWNTs with tin oxide (SnO_2) nanoparticles. The proposed uricase/MWNTs–SnO_2 biosensor showed excellent features of sensitivity and anti-interference, which enabled the monitoring of trace levels of uric acid in dialysate samples from rat striatum.[50]

MWNTs have also been modified with putrescine oxidase for detection of putrescine.[34] Polyamines play an essential role in the proliferation of normal and cancerous mammalian cells, and depletion results in inhibition of growth. By tests performed on plasma of cancerous mice, Rochette and co-workers, demonstrated that the detection of putrescine could be quickly carried out on mammalian plasma without previous purification. The biosensor was capable of efficiently monitoring the direct electroactivity of putrescine oxidase at the electrode surface, permitting the detection of putrescine, circumventing the interference of endogenous ascorbic and uric acids, which often complicate the analysis of important compounds in plasma. Putrescine was detected with a limit of 5 μM, and with a 20 times greater response compared to the most common interfering species (spermine, spermidine, cadaverine, and histamine).

Recently Veetil J. V. and Ye K. have extensively reviewed the current innovations for the development of CNT-based immunosensors, that could improve the sensitivity of the traditional ones.[42] Because of its high sensitivity, field effect transistor (FET) has been widely used in immunobiosensors. CNTs act as single conducting or network conducting channels and can be modified by the attachment of specific antibodies. Some of the mentioned papers are particularly interesting for the future development of cancer biomarkers analysis. For example, Park *et al.* immobilized monoclonal antibody against carcinoembryonic antigen (CEA) on the sidewalls of SWNT — FET using a carbonyldiimidazole-Tween 20 as linker. The antigen interaction was measured by the decrease of conductance.[31] Also Li C. *et al.*, described a novel FET immunosensor device for the prostate specific antigen (PSA), an oncological marker for the presence of prostate cancer. They reported complementary detection of PSA using n-type In_2O_3 nanowires and p-type carbon nanotubes. Anti-PSA antibodies were covalently attached to the SWNT by depositing 1-pyrenebutanoic acid succinimidyl ester on the nanotube surface (Fig. 1).[20]

After PSA incubation they observed enhanced conductance for NW devices and reduced conductance for SWNT. The complementary response in conductance can be understood as In_2O_3 NWs are n-type and SWNTs are p-type semiconductors.

The origin of the change of the device characteristics is that the chemical gating effect of PSA introduces carriers into In_2O_3 NWs, leading to enhanced conductance, while the PSA binding decreases the carrier concentration in nanotubes, thus reducing the conductance. In this way, the detection of PSA in solution has been demonstrated to be effective as low as 5 ng/mL, a level useful for clinical diagnosis of prostate cancer. Okuno J. and co-workers have fabricated a label-free electrochemical immunosensor using microelectrode arrays modified with SWNTs to detect total-PSA (T-PSA). The detection limit for T-PSA was determined as 0.25 ng/mL (cut-off limit of T-PSA between prostate hyperplasia and cancer is 4 ng/mL), opening promising clinical applications.[26] Recently, Briman M. and co-workers interestingly reported the use of a CNT-based sensor in the direct measurement of the detection of PSA protein in serum.[4]

A FET immunosensor was also fabricated for the detection of a human autoimmune disorder associated antigen. Detection of autoantibodies against the human auto antigen U1A is a common clinical assay for systemic lupus erythematosus (SLE). Chen *et al.* described a FET device with CNTs modified

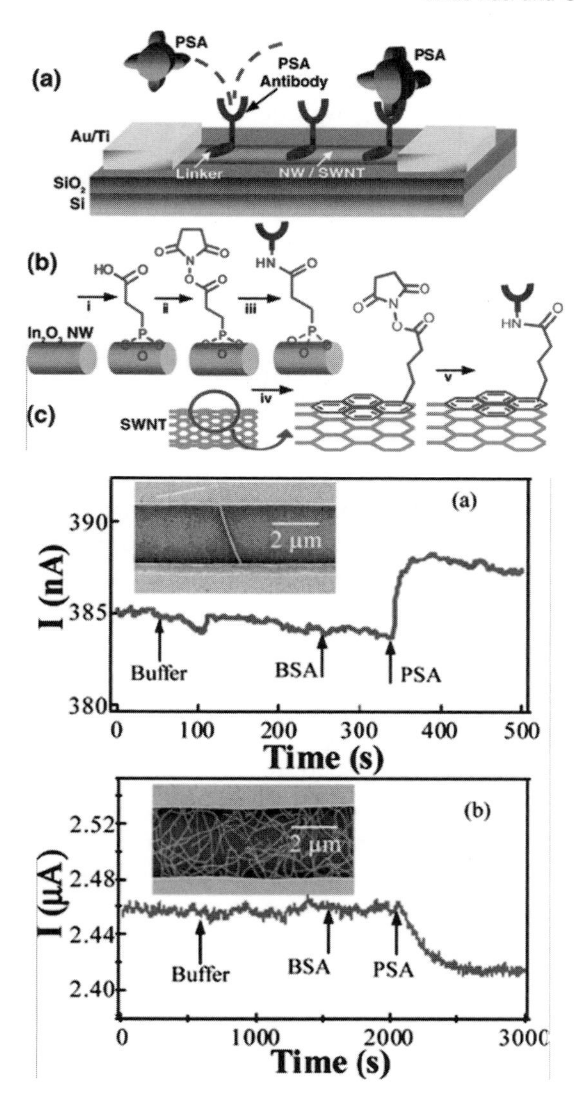

Fig. 1. (Top) (a) Schematic diagram of the nanosensor. PSA-ABs are anchored to the NW/SWNT surface and function as specific recognition groups for PSA binding. (b) Reaction sequence for the modification of In_2O_3 NW: (i) deposition of 3-phosphonopropionic acid; (ii) DCC and N-hydroxysuccinimideactivation; (iii) PSA-AB incubation. (c) Reaction sequence for the modification of SWNT: (iv) deposition of 1-pyrenebutanoic acid succinimidyl ester; (v) PSA-AB incubation. (Bottom) Current recorded over time for an individual In_2O_3 NW device (a) and a SWNT mat device (b) when sequentially exposed to buffer, BSA, and PSA. Insets: SEM images of respective devices. Reprinted with permission from Li C., *et al.*, *J. Am. Chem. Soc.*, **127**, 12484–12485.

with a recombinant human auto antigen U1A. The interaction of the antigen with specific monoclonal antibody resulted in a decrease in conductance.[6] All these findings open promising application for fast and reagent-less immunoassays, however, it is important to take under control, and to avoid potential non-specific binding that would give unpredictable measurements.

Deoxyribonucleic acid (DNA) biosensors utilize immobilized DNA as the biological recognition element and could be useful for the rapid diagnosis of genetic diseases, and for the detection of pathogenic biological species of clinical interest. Wanga S. G. *et al.*, designed a new DNA biosensors based on self-assembled MWNTs (Fig. 2).[45]

DNA hybridization was assessed with the changes in the voltammetric peak of an indicator of methylene blue. As shown in Fig. 2 the highest methylene blue signal was observed with the probe DNA/self-assembled MWNTs modified Au electrode before hybridization (a). Since the methylene blue has a strong affinity for free guanine bases, this indicates that the greatest amount of methylene blue accumulates on the electrode surface. Similar signal was achieved by the exposition to a non-complementary DNA oligonucleotides (b). In contrast, the lowest methylene blue signal was observed after incubation with the complementary target DNA sequence (d). Owing to the duplex formation, the signal from the methylene blue was suppressed. Compared to hybridization with the complementary target DNA sequence, a smaller decrease of the signals was observed when the probe DNA/self-assembled MWNTs modified Au electrode was hybridized with a three-base mismatch containing target DNA oligonucleotides (c). Furthermore the authors demonstrated that the DNA biosensors based on self-assembled MWNTs had a higher hybridization efficiency compared to those based on random MWNTs (data not shown). Those results demonstrate that the biosensors based on self-assembled MWNTs have a high selectivity of the hybridization detection.

Some authors have proposed CNT-biosensor to detect viruses. Takeda S. *et al.* have described a CNT-sensor capable of detecting Influenza hemagglutinin binding to a specific antibody immobilized on the biosensor.[38] Dastagir T. and co-workers have recently described the real time detection of a sequence of the Hepatitis C Virus (HCV) down to concentrations limit of 0.5 pM, using SWNT-FET devices functionalized with Peptide Nucleic Acid (PNA). Although the detection limit is much higher than what is needed for direct detection of HCV in untreated blood samples, the SWNT devices are label-free, fast, and reduce the use of PCR to 15 cycles from 50 cycles, opening interesting improvement for current detection methods.[9]

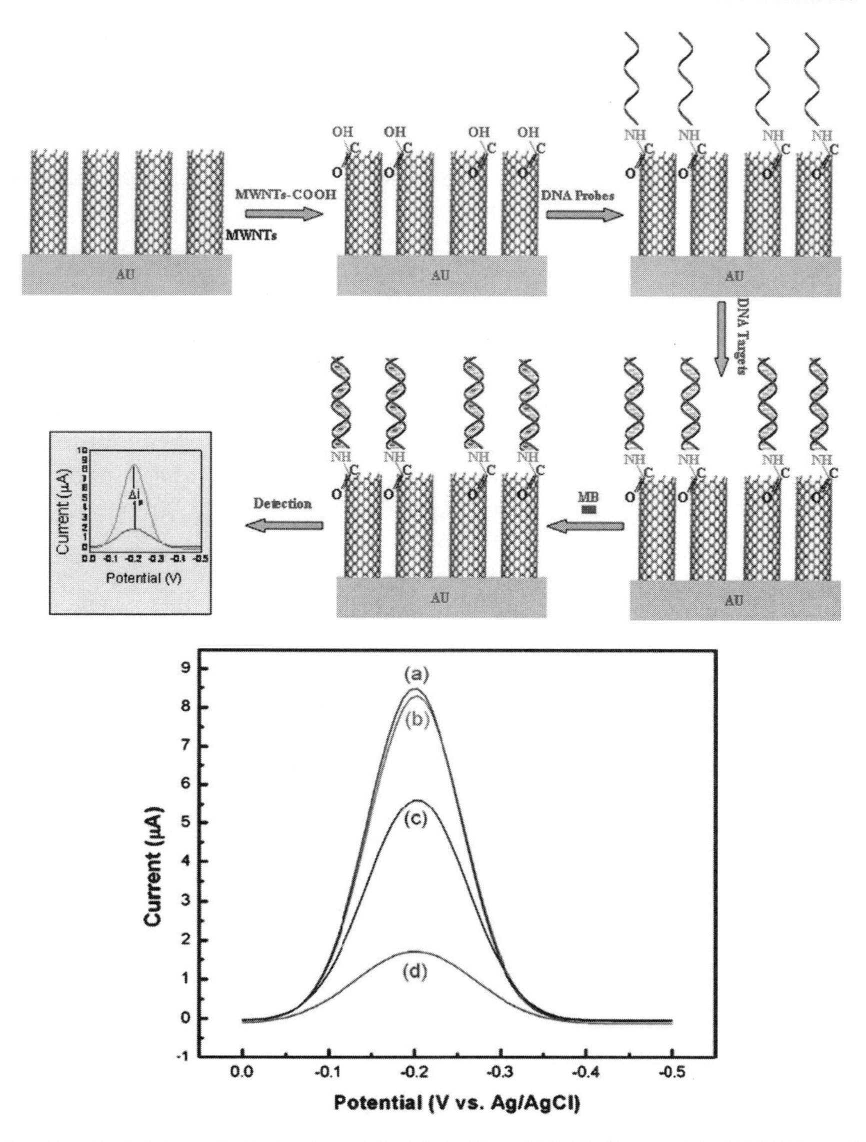

Fig. 2. (Top) Schematic illustration of the fabrication of DNA/self-assembled MWNTs modified gold (Au) electrodes and detection of target DNA sequences. (Bottom) Differential pulse voltammograms of probe DNA/self-assembled MWNTs modified Au electrodes before hybridization (a), after hybridization with non-complementary target DNA oligonucleotide sequence (b), with three-base mismatch containing target DNA oligonucleotides (c), and with complementary target DNA oligonucleotide sequence (d) at 50 mV/s scan rate in a 20 mM Tris–HCl buffer solution (pH 7.0). Reprinted with permission from Wanga S. G., *et al.*, *Biochem. Biophis. Res. Commun.* (2004) **325**, 1433–1437.

The development of implantable sensors remains a major challenge. Although, CNT-based biosensors offer new opportunities, at the present there are no *in vivo* applications described for these systems. Upon further optimizations CNT-based biosensor will compete with current analytical techniques providing additional advantages, such as reduced cost, minimal blood sample volume, direct electrical readout, and the ability to perform multiplexed detection for many biomarkers. An alternative approach to electrochemical sensing is through the use of optical methods. CNTs provide many characteristics that are suitable for detection in the infrared or near infrared optical regions; furthermore they can be easily detected by chemical functionalization. The optical window between 900 and 1300 nm has important biomedical applications because of greater penetration depths of light and small autofluorescent background without harming tissues and blood. Taking advantage of this features, Barone P. W. and colleagues have used modified SWNT to measure b-D-glucose concentrations, monitoring the NIR fluorescence from the SWNT that changed upon binding of glucose.[2]

Recently, the same research group together with other authors demonstrated for the first time the multifuntionality of single-walled carbon nanotube/iron oxide nanoparticle complexes as dual magnetic and fluorescent imaging agents. By encapsulation with DNA, the SWNT/iron oxide nanoparticle complexes were individually dispersed in aqueous solution, and were more easily introduced into a biological environment. They showed 2-D *in vitro* images of murine macrophage cells containing these nanostructures, using magnetic resonance and near-infrared mapping. These results suggest that these complexes could be used to assess tissue or probe individual cells of interest.[7] Further functionalized SWNT/iron oxide nanoparticle, with monoclonal antibodies to target specific receptor sites, could be used to provide molecular-level contrast and biosensing. Finally, the potential exists for these complexes to achieve phototherapy and hyperthermia effects in cells and tissue through NIR laser radiation, and high-speed rotation of the nanomaterials upon application of an external magnetic field modulated at a high frequency.

CNTs can also be modified with radiotracers for gamma scintigraphy. In one case, indium (^{111}In) was covalently bound to SWNTs and administered to BALB/c mice to measure the biodistribution of nanotubes.[36] Therefore, heavy elements functionalized to CNTs could serve as X-ray contrast agents. The ability to track implanted cells, and to monitor the progress of tissue formation *in vivo* and non-invasively, is important especially in tissue-engineered constructs of clinically relevant sizes. Labelling implanted cells would help in evaluating the viability of the engineered tissue, but, would also help in the understanding of the biodistribution and migration pathways of transplanted cells.[12] However, contrasting *in vivo* agents need to have good biocompatibility, high contrasting ability, and stability. CNTs posses many properties desirable for optical detection. CNTs remained in the cells during repeated cells divisions, suggesting that such probes could be used for studying cell proliferation and differentiation. However, complementary studies in biocompatibility will make clear potential side affects for these applications.

Numerous procedures in cell therapy and tissue engineering suffer for a difficulty in monitoring the progress of tissue regeneration or cell behaviour; non invasive methods to do this would be extremely helpful. The goal of tissue engineering is to replace diseased or damaged tissue with biologic substitutes that can restore and maintain normal functions. Important advances in the areas of cell and organ transplantation, as well as advances in materials science and engineering, have aided in the continuing development of tissue engineering and regenerative medicine. Indeed, CNTs are emerging as promising material that could have significant impact in this medical field.[12] In particular, CNTs have the ability to serve as multifunctional structural materials providing the initial structural support needed for newly created tissue scaffolds.

It is particularly interesting that CNTs appear to be good substrates for neuron growth. In fact, CNTs represent a scaffold composed of small fibres or tubes that have diameters similar to those of neural processes such as dendrites. Lovat V. and colleagues succeeded in having a long term and stable retention of films of CNTs on glass, not affected after eight days of incubation in the culture medium (Fig. 3(A)). Fixed hippocampal neurons were analyzed by SEM after 8–10 days *in vitro*. As shown in Figs. 3(B) and (C), neurons grew attached to the purified CNTs and extended several neurites, which grew to various distances.

Zanello L. *et al.* explored the use of CNTs as suitable scaffold materials for osteoblast proliferation and bone formation.[49] They found that osteoblasts grow and produce mineralized bone when cultured on electrically neutral CNTs. When cultured on CNTs, ROS 17/2.8 cells retained electrical properties necessary for adequate secretion of bone materials. CNTs showed biocompatibility with osteoblast cells, and appeared to modulate the cell phenotype (Fig. 4). Moreover, CNTs sustained osteoblast growth and bone formation,

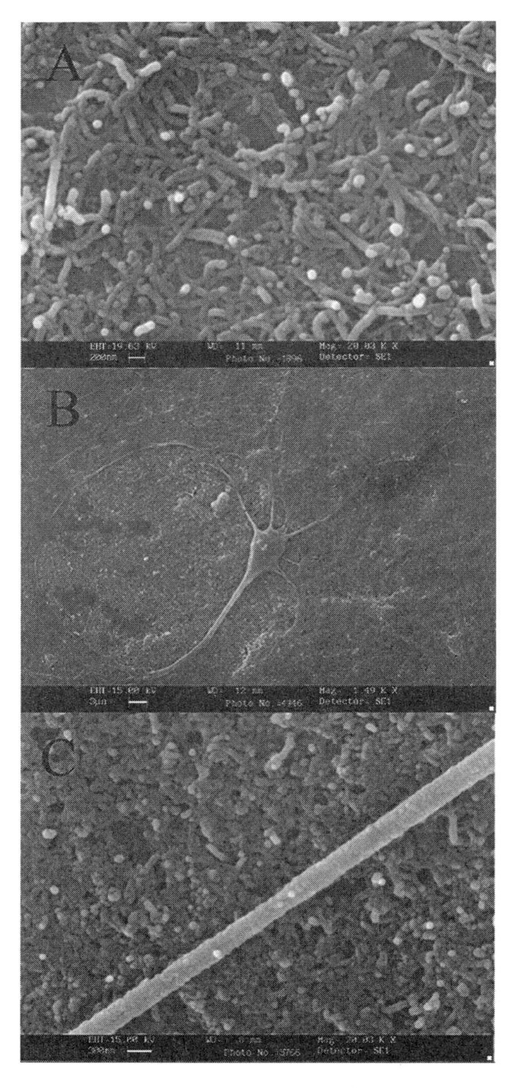

Fig. 3. Purified multiwalled carbon nanotubes (MWNT) layered on glass are permissive substrates for neuron adhesion and survival. (A) Micrographs taken by the scanning electron microscope showing the retention on glass of MWNT films after an eight-day test in culturing conditions. (B) Neonatal hippocampal neuron growing on dispersed MWNT after eight days in culture. The surface structure, composed of films of MWNT and peptide-free glass, allows neuron adhesion. Dendrites and axons extend across MWNT, glia cells, and glass. The relationship between dendrite and MWNT is very clear in the image in (C), where a neurite is traveling in close contact to carbon nanotubes. Reprinted with permission from Lovat V. *et al.*, *Nano Lett.* (2005) **5**, 1107–1110.

and so, thus represent a potential technological advance in the field of bone bioengineering. CNTs have shown capacity to be used in chemical and biological sensing, cellular imaging and matrix engineering, thus they are candidate as innovative material to develop multivalent platforms for biomedical application.

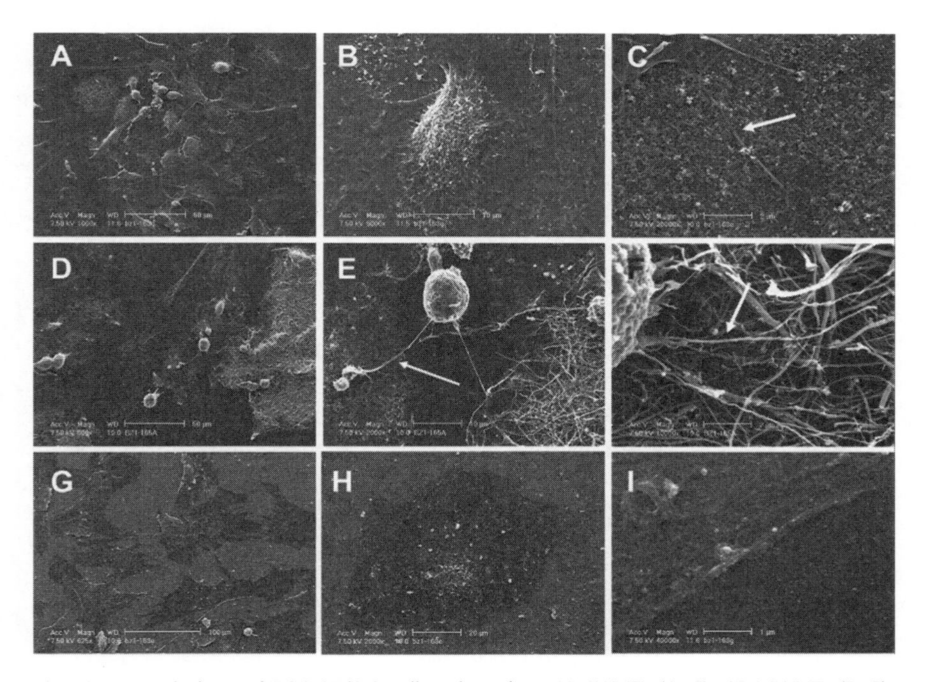

Fig. 4. Morphology of ROS 17/2.8 cells cultured on AP-SWNTs (A–C), AP-MWNTs (D–F), and control cultures on glass cover slips (G–I), as seen with SEM. (A) Osteoblast colony on AP-SWNTs. (B) A flat cell body of a single cell extends over almost the entire field of observation; the cell nucleus protrudes in the center. (C) Tape-like cytoplasmic prolongations (arrow) extend from the flat body of a ROS 17/2.8 cell (a portion of it shown at the left upper corner of the picture) on an evenly distributed AP-SWNT substrate. (D) Osteoblast colony on AP-MWNTs. The nanotubes aggregate unevenly in areas of the glass surface (notice bundles on CNTs on the right). (E) Image of a single ROS 17/2.8 cell on AP-MWNTs. A round single-cell body extends thin neurite-like cytoplasmic prolongations (arrow) that reach the nanotube bundles. (F) SEM micrographs at higher magnification show a detail of long threadlike cytoplasmic prolongations (arrow) that extend from the round body of a single ROS 17/2.8 cell (partially seen on the left, upper corner), interweave with, and reach individual AP-MWNTs. (G) ROS 17/2.8 cell colony cultured on glass. (H) Image of a single cell obtained at higher magnification. (I) Detail of a portion of the cell cytoplasm (covering the left upper half of the picture) in contact with the glass surface; no cytoplasmic prolongations are observed. Reprinted with permission from Zanello L. P. *et al.*, *Nano Lett.* (2006) **6**, 562–567.

Overall the experience here reported indicate an upcoming relevant role for CNTs in the diagnostic field. Point of strength are the possibility to have a reagent-less test, particularly attractive for cost reduction, safeness and simplicity of handling and the possibility to develop a multitest platform. This is particularly exciting because multitesting is the real actual need in *in vitro diagnosis* (IVD). Gene expression, proteomic, metabolomic and peptidomic profiling is the requirement for the next generation medicine, linking diagnosis and therapy. The data will be more and more available, thanks to the improvement of the high throughput techniques (mass spectrometry, data mining etc.). However the existing technologies are mostly devoted to research and do not allow for friendly and rapid carrying out of tests and real time availability of results, as required in a clinic setting. The critical points for IVD seem to be those expected for any test: validation, reproducibility and robustness, along with cost control; but this belong to the odds that any research imply.

To develop implantable biosensors the way may be longer, the problem of biocompatibility and toxicity should be addressed and the risk/benefit ratio exactly determined. Overall the benefits coming by the possibility to routinely use biosensors, such as those described, and the proof of concept provided *in vitro* are exciting enough to encourage the development. The use for *in vivo* tracking and tissue engineering require significant financing efforts mostly for the development of dedicated preclinical models. This, however, may represent an additional positive side effect of CNTs technology development.

3.2. Application in Drug, Peptide and Gene Delivery

The technologies for the development of new methods to deliver drugs, peptides or nucleic acids, in a fast, durable, in time and safe ways, are constantly improving to optimize pharmacokinetic and targeting with lower side effects.

In the last years nanotechnology and nanofabrication have offered significant advances in the field of drug delivery. Both nano and micro scale systems have been extremely important in developing various clinically useful drug delivery systems, ranging from truly nanosystems (e.g. drug–polymer conjugates and polymer micelles) to micro particles up to 100 μm.[31]

CNTs represent a new alternative for transporting therapeutic molecules; with high surface area to volume ratio and ease to be chemically functionalised. CNTs suggest potential applications in targeted drug delivery and gene transfection. Due to the chemical reactivity of the surface of CNTs, several functional groups can be located at the ends or around the side-wall of the CNTs.[1] The ends of the nanotubes could be functionalised with a antibody to target the nanotube to a particular receptor on a cell, at the same time the side walls could be functionalized with a drug attached via a biodegradable linker.

CNTs can be functionalized with bioactive proteins for help in crossing the cellular membrane. Bianco *et al.* demonstrated how the functionalisation of carbon nanotubes, other that improving their solubility and biocompatibility, trasformors CNTs in platforms for biomedical application.[3] CNTs functionalised with amino group (f-CNT) have been covalently linked to amino acids, fluorescent probe and bioactive peptides. The aqueous solubility and cationic surface, permit f-CNT to cross the plasma membrane and distribute throughout the cellular compartments. Using a chemical ligation approach, a molecule of fluorescein isothiocyanate (FITC) or a fluorescent peptide were attached to the amino functions of carbon nanotubes.[27,28] The modified peptide belonged to the α subunit of the Gs protein with properties to increase the agonist affinity for the β-adrenergic receptor. In contrast of the weak cellular uptake of the free peptide, an experiment of treatments on Hela cell, demonstrated that f-CNTs conjugated with FITC alone penetrate in the cell, localising mainly in the cytoplasm, but the peptide-CNTs rapidly traslocate into the nucleus as shown by fluorescence microscopy (Fig. 5).

Other groups demonstrated that biotin functionalized CNTs bounded to fluorescent dyes were capable to transport fluorescent streptavidin inside HL-60 cells.[15] In another study, Kam N. W. S. and Dai H., reported that SWNTs are intracellular transporters for various type of proteins (streptavidin, protein A, serum bovine albumin and cytochrome C-cyt-c) in different kind of cell lines (Hela, NIH-3T3 fibroblast, HL-60 and Jurkat).[14] After SWNTs-mediated delivery of cyt-c in NIH-3T3 cells, they observed that chloroqhine treatment permitted endosomes release of the internalized molecules by consequently induced cyt-c dependent apoptosis, demonstrating the proof of the concept of *in vitro* biological activity for protein delivered by SWNTs (Fig. 6).

In a comparative study between administration of different kinds of nanoparticle based delivery systems in rats, Venkatesan *et al.* showed that

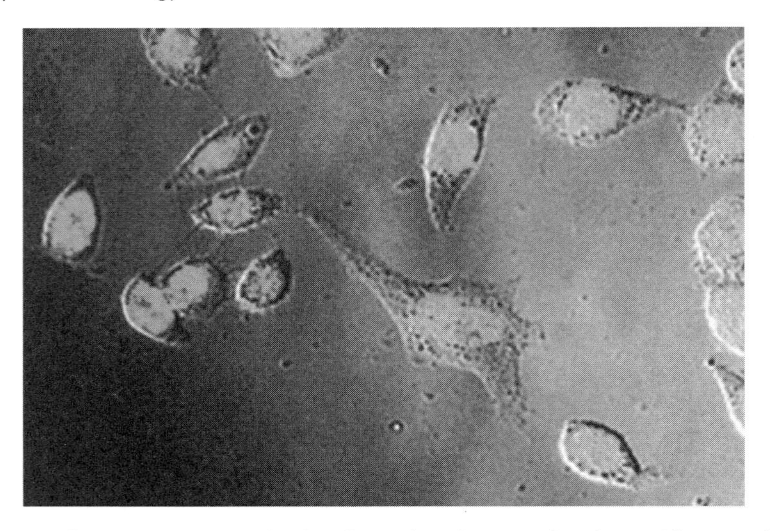

Fig. 5. Epifluorescence image of Hela cells incubated at 37°C for 1 hour with 5 μM of fluorescent peptide f-CNTs. Reprinted with permission from Bianco *et al., Chem. Commun.* (2005) 571–577.

CNTs were the best system to improve bioavailability of erythropoietin (EPO).[43] Because of the denaturation of EPO in the gastric tract, the system included an intestinal enzyme inhibitor, and an absorption enhancer, demonstrating that CNTs could offer a new way of EPO administration.

Other authors functionalised CNTs with Amphotericin B (AmB), to test the antifungal effect of this compound. Compared to the free AmB, the AmB-MWNTs reduce the cellular toxic effect of AmB; the activities of this antibiotic were not affected by the conjugation with the CNT; in addition, a decrease in the Minimum Inhibitory Concentration (MIC) was achieved by AmB-SWNTs and AmB-MWNTs, compared to the free antibiotic against *Candida albicans, Candida parapsilosis and Cryptococcus neoformans*.[48] The functionalization of CNTs with antigens and their ability to induce an antigenic response *in vivo*, were studied to propose CNTs as carriers for the vaccination peptides delivery.

CNTs were functionalised with an epitope of the VP1 protein of foot-and-mouth disease virus (FMDV), selected by the authors as the epitope that induces neutralizing protective antibody against the virus.[28,29] The viral-peptide-CNTs were recognised by the specific antibodies as well as the free peptide, while the CNTs devoid of the peptide were not recognised. The *in vivo* data demonstrated that the immunization of mice with

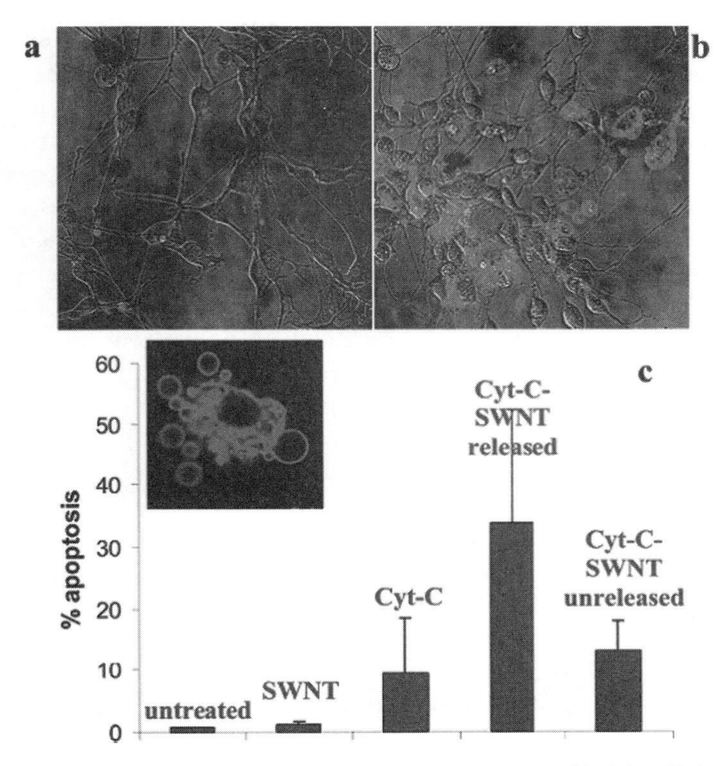

Fig. 6. Apoptosis induction by cytochrome c cargos transported inside cells by SWNTs. (a) Confocal image of NIH-3T3 cells after 3 h incubation in 50 μM cyt-c alone and 20 min staining by Annexin V-FITC (green fluorescent). (b) 3T3 cell after incubation in 50 μM cyt-c-SWNTs in the presence of 100 μM chloroquine and after Annexin V-FITC staining. (c) Cell cytometry data of the percentages of cells undergoing early stage apoptosis (as stained by Annexin V-FITC) after exposure to 100 μM chloroquine only ("untreated"), $\mathrm{SWNT} + 100\,\mu\mathrm{M}$ of chloroquine, 10 μM cyt-c + 100 μM chloroquine, 10 μM $\mathrm{cyt-c-SWNT}+100\,\mu\mathrm{M}$ chloroquine, and cyt-c-SWNT without chloroquine. The inset shows a representative confocal image of the blebbing of the cellular membrane (stained by Annexin V-FITC) as the cell undergoes apoptosis. Reprinted with permission from Kam N. W. S. and Dai H., *J. Am. Chem. Soc.* (2005) **127**, 6021–6026.

viral-peptide-CNTs, compared to the free peptide enhanced anti-FMVD peptide antibody responses, with production of specific antibodies not directed to CNTs or to the peptide linker. Due to their low antigenicity CNTs are interesting and promising carrier systems for vaccines antigens delivery.

CNTs have also been conjugated with deoxyribonucleic (DNA) acid and ribonucleid acid (RNA), with cleavable linking molecules, and delivered to target cells for gene transfection.

There are many approaches for delivering genetic materials to cells, including the use of viral (retroviral, lentiviral, or adenoviral) vectors, and non-viral methods such as cationic organic molecules and polymers. While viral methods have high transfection efficiencies, they have intrinsic risk factors due to undesired immunogenic pro-inflammatory and pro-oncogenic properties.[41] Non-viral vectors were also proposed for gene delivering as liposomes, cationic lipids, polymers and micro- and nano-particles of variable source, but the critical point of this method is the pharmacokinetic and the low levels of gene expression. Furthermore, while CNTs can be functionalized to attach either electrostatically or covalently to DNA and RNA, the remaining unfunctionalized and hydrophobic portions of the nanotubes, can be attracted to the hydrophobic regions of the cells, or can also be functionalized to facilitate DNA transfection of cells with components capable of penetrating human and murine cell types. The interaction of CNTs with single stranded DNA (ssDNA) have been studied. Zheng M. and co-workers demonstrated that SWCNTs have high affinity for ssDNA, presumably by hydrophobic interactions.[53] Bianco A. and co-workers examined the potential of functionalised CNTs to form supramolecular complexes with plasmid DNA.[3,30,37] Due to the only electrostatic interactions, they were able to form a complex between a pCMV-bgal plasmid and f-CNTs, The resulting supramolecular assemblies were observed using Trasmission Electron Microscopy (TEM) (Fig. 7).

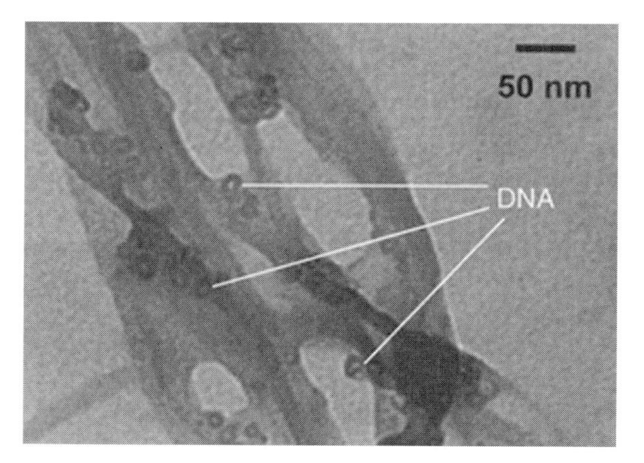

Fig. 7. Trasmission Electron Microscopy image of f-CNT:DNA complexes. Reprinted with permission from Bianco *et al., Chem. Commun.* (2005) 571–577.

Moreover, they were able to monitor the expression of β-galactosidase gene inside the cells, demonstrating that the gene expression induced by the f-CNTs-DNA complexes was 5 to 10 times higher than that of DNA alone. However, preliminary data showed that this new gene delivery system was less efficient for *in vitro* transfection than the lipid one. However, these promising results and the innovative research in CNT technology, actually offer the possibility to develop new strategies for the application in gene therapy and genetic vaccination.

In the area of drug delivery for cancer therapy, CNTs have primarily been used for transporting DNA cargoes into the cell and for thermal ablation therapy (see Section 3.3 of this chapter). Exposing the DNA-nanotube containing cells to near infrared (NIR) light, caused endosomal rupture, unloading of the DNA from the CNTs, and translocation into the nucleus.[16]

Another strategy to regulate gene expression is gene silencing by siRNA. Kam N. W. S. *et al.* functionalized CNTs via cleavable disulfide bonds to enhance intracellular delivery of siRNA and potentiate gene silencing.[16] The non-covalent adsorbation of phospholipid molecules in presence of poly(ethylene glycol), made a stable aqueous suspensions of SWNT; the terminal amine or maleimide could be used to conjugate with a wide range of biological molecules, including DNA and proteins. The disulfide linkage between the attached molecules and SWNT sidewall, is cleaved by lysosomal enzymes for releasing into cytosol. With this system the authors were able to dowregulate the expression of the lamin A/C gene, delivering specific siRNA with higher efficacy of lipofectamine based transfection.

More recently, Liu Z. and co-workers have shown that nanotubes are capable to deliver siRNA to human T cells and peripheral blood mononuclear cells (PBMCs). They achieved specific knockdown of cell-surface CXCR4 and CD4 receptors by SWNT-mediated specific siRNA delivery into CEM cells, with high efficiency compare to transfection by lipofectamine system.[22] Due to these results, CNTs based RNA-interference exceed those of several existing nonviral transfection agents, and in suggesting promising biological and medical application, CNTs have shown to easily penetrate into the cell, to promote proteins and nucleic acids translocation in cellular compartments, to allow gene expression and peptide-specific antigen stimulation.

The technology described is convincing and advanced enough to consider CNTs as promising carrier for drug delivery, gene therapy and new vaccination strategies, with prospects for coupling these features to biosensors and implants. The major problems for CNTs drug delivery are those described for implantable sensors. Supplementary studies on the cellular toxicity and body distribution and elimination are necessary to apply this new technology for the improvement of human health.

However, meanwhile the final answer about toxicity and biocompatibility will be available some basic distinctions can drive the strategy for the development of these devices. Among those the possibility to remove them from the body, the amount of CNTs to administer and the duration of the treatment, related to the severity of the disease and to the toxicity of the alternative existing therapy.

3.3. Application in Cancer Diagnosis and Therapy

Owing to its interdisciplinary features, the research in the field of nanotechnology for the treatment of cancer involves the efforts of biologist and chemist researchers, oncologists, immunologists, electronic and mechanical engineers. During the last few years, the intellectual and monetary investments, from the scientific and economic communities of the greater nations in this field, reflect a great expectation toward the potential application of nanotechnology for a drastic reduction of cancer related deaths and for the development of personalized therapies.

The potential of cancer nanotechnology is in the ability to engineer vehicles with unique therapeutic properties that, because of their small size, can penetrate tumors with a high-level specificity. By delivering genes, proteins or drugs directly to the cells, CNTs might allow to quickly detect and treat cancer at the early stage.[11,18]

Thus, CNTs could be applied to tools for biosensing, imaging and delivery, capable of selectively target the tumor and simultaneously be therapeutic. These rapidly emerging tools have brought fundamental advantages for cancer detection and treatment applications. For example, CNT-based sensor elements for selective multiplexed could sense cancer markers without the need of probe labelling (see Section 3.1 of this chapter). Furthermore, as already mentioned above, besides heterogeneous functionalization, CNTs could provide localized delivery of therapeutic agents.

CNTs absorb near-infrared (NIR) light at wavelengths that are optically transparent to native tissue. Irradiation with 880 nm laser pulses can induce local heating of SWNTs *in vitro*; this effect can induce the releasing of the therapeutic molecules without harming cells or CNTs, and when internalized within a cancer cell, with sufficient heating can kill the cell.[17] CNTs could be also functionalized by adsorbing different molecules, antibodies or antigens so that they may specifically target tumor cells. Kam *et al.* used folic acid (FA) and fluorescent tag conjugated SWNTs, to target tumoral cells highly expressing FA receptors cells *in vitro*, demonstrating that, SWNTs can be selectively internalized into cancer cells with specific tumor markers. SWNTs were actively internalized and then identified through the Fitc-tag by confocal microscopy analysis. In addition, NIR *in vitro* radiation of the nanotubes, could selectively activate cell death without harming normal cells (Fig. 8).

Zhang Z. *et al.*[52] demonstrated that CNTs carrying interfering RNA (siRNA) can rapidly enter tumor cells, to exert the RNA interference on target gene expression (see also Section 3.1 of this chapter). They used siRNA that specifically targeted murine telomerase reverse transcriptase coupled to SWNTs by the functional group — CONH-(CH2)6-NH3. The activation of telomerase is critical for immortalization, and it is detected in numerous malignant tumors, but not in most of normal cells. Hence, inhibition of telomerase activity is a major goal in targeted cancer therapy. The delivery of telomerase-siRNA via CNTs silenced the target gene at the mRNA and protein levels, but also inhibited the proliferation of cancer cells *in vitro*, and suppressed tumor growth in mouse models, upon intralesional injection (Fig. 9). The use of CNTs as a vehicle for delivery of siRNA in cancer cells, presents great promises.

Liu Z. *et al.* have recently described the *in vivo* distribution of radio-labelled SWNTs in mice by positron emission tomography (PET). The unique one-dimensional shape and flexible structure of SWNTs enables a polyvalence effect and enhances tumor binding affinity. Efficient targeting of integrin positive tumour in mice was achieved with SWNTs coated with polyethylene-glycol (PEG) chains, linked to an arginine–glycine–aspartic acid (RGD) peptide. They did not observe toxicity or negative health effects. However, SWNTs functionalized with PEG showed elevated stability *in vivo*, and moreover, relatively slow excretion and high accumulation in the liver of SWNTs was observed.[23]

Other authors evaluated the biodistribution of antibody-functionalized radiolabeled CNTs, developed as a platform, with attached antibody, metal-ion chelate, and fluorescent chromophoremoieties, to respectively effect specific targeting, to carry and deliver a radiometal-ion, and to report location.[24] Although the constructs demonstrated to perform their functions in targeting cancerous cells, it's not certain if the accumulation in liver and kidney can be acceptable for clinical applications. Moreover, the covalent attachment of antibody molecules to the CNTs dramatically altered the kidney biodistribution and pharmacokinetics. Supplementary investigation, to predict modifications that could affect the pharmacokinetic profile and to understand the

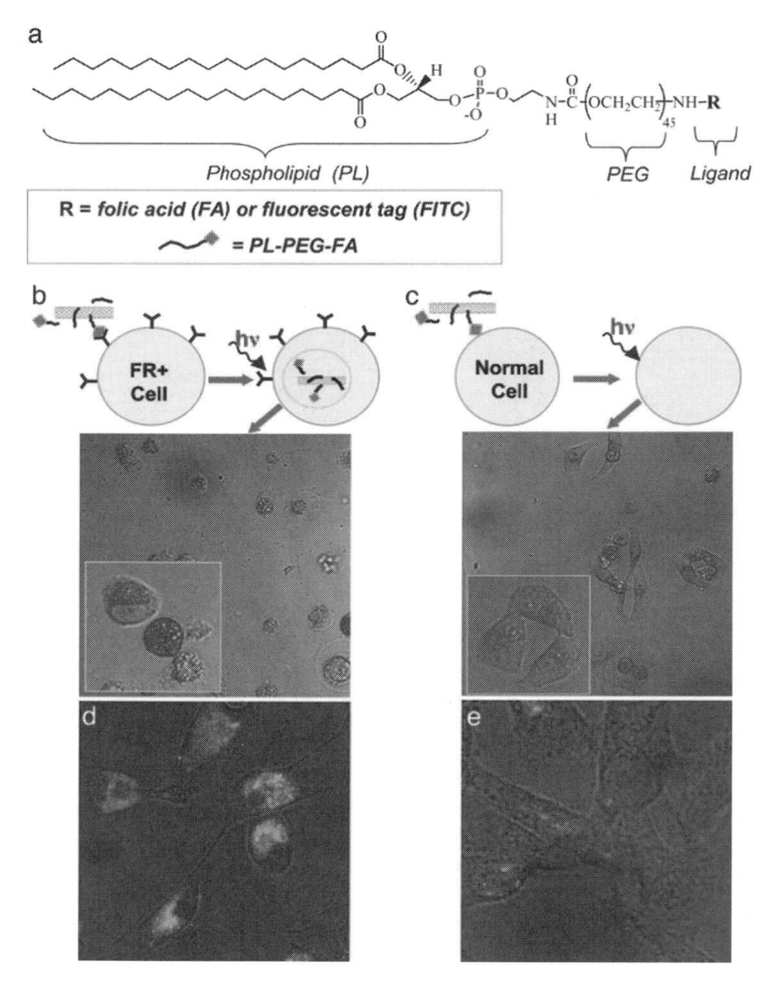

Fig. 8. Selective targeting and killing of cancer cells. (a) Chemical structure of PL-PEG-FA and PL-PEG-FITC synthesized by conjugating PL-PEG-NH2 with FA or FITC, for solubilizing individual SWNTs. (b) (Upper) Schematic of selective internalization of PL-PEG-FA-SWNTs into folate-overexpressing (FR+) cells via receptor binding and then NIR 808-nm laser radiation. (Lower) Image showing death of FR+ cells with rounded cell morphology after the laser radiation (808-nm at 1.4 W_cm2 for 2 min). (Inset) Higher magnification image shows details of the killed cells. (c) (Upper) Schematic of no internalization of PL-PEG-FA-SWNTs into normal cells without available FRs. (Lower) Image showing normal cells with no internalized SWNTs are unharmed by the same laser radiation condition as in (b). (Inset) Higher magnification image shows a live normal cell in stretched shape. (d) Confocal image of FR+ cells after incubation in a solution of SWNTs with two cargoes (PL-PEG-FA and PL-PEG-FITC). The strong green FITC fluorescence inside cells confirms the SWNT uptake with FA and FITC cargoes. (e) There is little green fluorescence inside cells, confirming little uptake of SWNTs with FA and FITC cargoes in normal cells (magnification: ×20). Reprinted with permission from Kam N. W. S. *et al.*, *Proc. Natl. Acad. Sci.* (2005) **102**, 11600–11605.

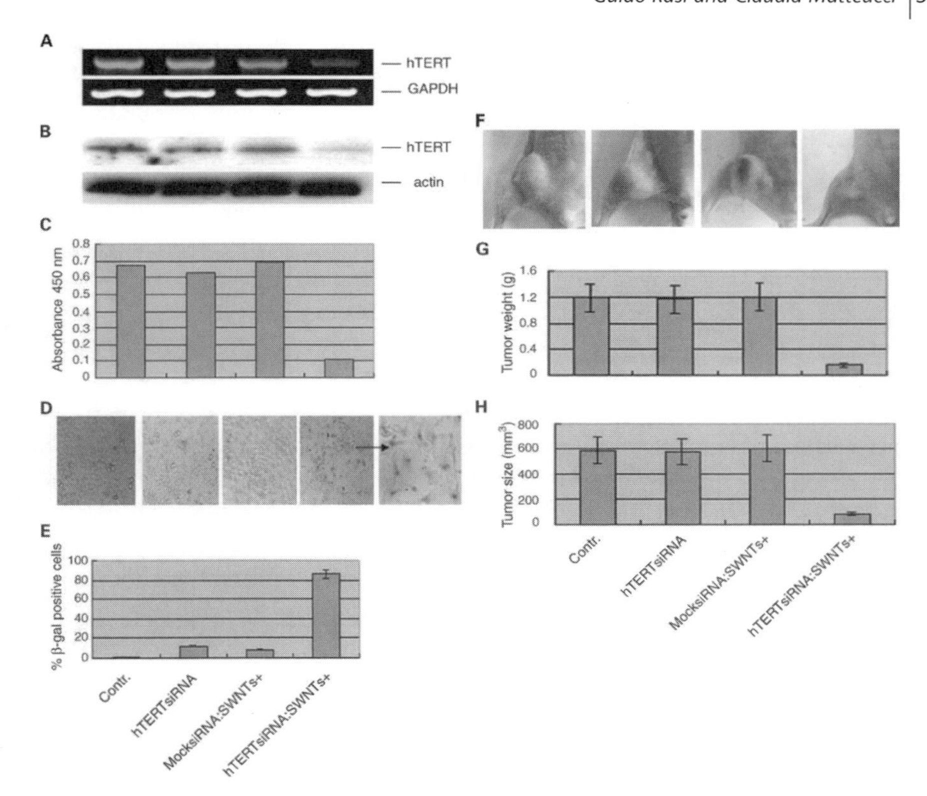

Fig. 9. Injection of hTERTsiRNA: SWNT + complexes suppresses HeLa xenograft growth. A to C, specific gene silencing induced by hTERTsiRNA: SWNT+ complexes in human HeLa cells. The level of hTERT transcripts (A), hTERT protein (B), and telomerase activity (C) was assayed in untreated control HeLa cells (lane 1) and after treatment of HeLa for 24 hours with 2 nmol/L of mTERTsiRNA alone (lane 2), mock siRNA:SWNTs+ (lane 3), and hTERTsiRNA: SWNTs+ (lane 4). D and E, *in vitro* suppression of human+ HeLa cell proliferation and induction of senescence by hTERTsiRNA: SWNT+ complexes; untreated control HeLa cells (left) and after treatment of HeLa for eight days with 2 nmol/L of mTERTsiRNA alone, mock siRNA: SWNTs+, and hTERTsiRNA: SWNTs + (middle right and right). E, the number of blue-stained cells was counted in at least 10 fields at ×400 magnification and expressed as a percentage of total cell number. F to H, injection of hTERTsiRNA: SWNTs+ retarded HeLa tumor growth. HeLa cells (2×10^6 per mouse) were injected s.c into right leg of nude mouse (five mice per group). After one week, when diameter of tumors was f2.5 mm, hTERTsiRNA: SWNTs or siRNA alone or mock siRNA: SWNT+ complexes were injected into the tumor at multiple points (100 AL of 10 nmol/L siRNA per mouse). F, one representative of per group of control animals (left), hTERTsiRNA alone (middle left), mock siRNA: SWNT + (middle right), and hTERTsiRNA: SWNTs (right). G, columns, mean tumor size; bars, SE. H, columns, mean tumor weight; bars, SE. Zhang Z. *et al.*, *Clin. Cancer Res.*, **12**, 4933–4939.

reasons for accumulation and clearance of CNTs in normal tissue, are crucial for further development.

Discovery in CNTs applications is in progress, with important implications for the management of cancer patients in the near future. The development of integrated platforms could allow simultaneously detection, imaging and therapy of cancer. Utilizing a "tumor-directed" therapy, agents can be achieved in the site of action with minimal systemic exposure, reducing toxicity and increasing its availability. Application of cancer diagnosis and therapy in nanoscale component or devices, offer the opportunity to introduce promising novel approaches to cancer treatment, delivering multiple drugs in a timed manner and at different locations in the body.

The plurality of functions potentially performed by the CNTs perfectly fits with the complex strategy required for cancer diagnosis and treatment. Moreover the cost benefit in term of toxicity is more favourable because of the severity of the disease, and the length of treatment should be necessarily short. Additionally a variety of preclinical cancer models are available, suitable also for toxicity, biodistribution, biocompatibility studies. Taken together, these considerations make cancer management the most likely field of application for CNTs technology in the near future.

3.4. Conclusions

A large number of potential biological and medical applications for carbon nanotubes based on their particular small, plastic, traceable properties have been establish.

However, CNT-based biomedical applications are still in the nascent stage, and there are still many challenges to be overcome for the successful commercialization of the concepts. Future applications of carbon nanotubes are desirable beside possible toxic effects. Recent research studies demonstrated relatively low toxicity of CNTs supporting the promising utilization of this new technology. In every therapy cost benefit evaluation is required in terms of efficacy, safety and costs. The toxicity of nanotubes can be compensated by the reduction of drug toxicity (as shown) or vice versa, the benefit of the delivery in terms of drug/cost reduction and bioavailability might be impaired by the CNTs toxicity.

The use in tissue engineering seems promising but far, the use as implantable biosensor and drug delivery tools might belong to the near future. The use as IVD belong to the present.

Bibliography

1. Banerjee, S., Hemraj-Benny, T. and Wong, S. S. (2005) Covalent surface chemistry of single-walled carbon nanotubes, *Adv. Mater.*, **17**, 17–29.
2. Barone, P. W., Baik, S., Heller, D. A. and Strano, M. S. (2007) Near-infrared optical sensors based on single-walled carbon nanotubes, *Nature Mat.*, **4**, 86–92.
3. Bianco, A., Kostarelos, K., Partidos, C. D. and Prato, M. (2005) Biomedical applications of functionalised carbon nanotubes, *Chem. Commun.*, **5**, 571–577.
4. Briman, M., Artukovic, E., Zhang, L., Chia, D., Goodglick, L. and Gruner, G. (2007) Direct electronic detection of prostate-specific antigen in serum, *Small*, **3**, 758–62.
5. Bockrath, M., Cobden, D. H., McEuen, P. L., Chopra, N. G., Zettl, A., Thess, A. and Smalley, R. E., (1997) Single-electron transport in ropes of carbon nanotubes, *Science*, **275**, 1922–1925.
6. Chen, R. J., Bangsaruntip, S., Drouvalakis, K. A., Kam, N. W., Shim, M., Li, Y., Kim, W., Utz, P. J. and Dai, H. (2003) Noncovalent functionalization of carbon nanotubes for highly specific electronic biosensors, *Proc. Natl. Acad. Sci.*, **100**, 4984–4989.
7. Choi, J. H., Nguyen, F. T., Barone, P. W., Heller, D. A., Moll, A. E., Patel, D., Boppart, S. A. and Strano, M. S. (2007) Multimodal biomedical imaging with asymmetric single-walled carbon nanotube/ion oxide nanoparticle, *Nano Lett.*, **7**, 861–867.
8. Cui, H. F., Ye, J. S., Zhang, W. D., Li, C. M., Luong, J. H. and Sheu, F. S. (2007) Selective and sensitive electrochemical detection of glucose in neutral solution using platinum-lead alloy nanoparticle/carbon nanotube nanocomposites, *Anal. Chim. Acta.*, **594**, 175–83.
9. Dastagir, T., Forzani, E. S., Zhang, R., Amlani, I., Nagahara, L. A., Tsuib, R. and Tao, N. (2007) Electrical detection of hepatitis C virus RNA on single wall carbon nanotube-field effect transistors, *Analyst*, **132**, 738–740.
10. European Commission (2005) European Technology Platform on NanoMedicine — Nanotechnology for Health, *Office for Official Publications of the European Communities Luxembourg*, http://www.cordis.lu/nanotechnology/nanomedicine. htm pp. 1–37
11. Ferrari, M. (2005) Cancer nanotechnology: Opportunities and challenges, *Nat. Rev. Cancer*, **5**, 161–171.
12. Harrison, B. S. and Atala, A. (2007) Carbon nanotube applications in tissue engineering, *Biomaterials*, **28**, 344–353.
13. Iijima, S. (1991) Helical microtubules of graphitic carbon, *Nature*, **354**, 56–58.
14. Kam, N. W. S. and Dai, H. (2005) Carbon nanotubes as intracellular protein transporters: Generality and biological functionality, *J. Am. Chem. Soc.*, **127**, 6021–6026.
15. Kam, N. W. S., Jessop, T. C., Wender, P. A. and Dai, H. (2004) Nanotube molecular transporters: Internalization of carbon nanotube–protein conjugates into mammalian cells, *J. Am. Chem. Soc.*, **126**, 6850–6851.
16. Kam, N. W. S., Liu, Z. and Dai, H. (2005) Functionalization of carbon nanotubes via cleavable disulfide bonds for efficient intracellular delivery of siRNA and potent gene silencing, *J. Am. Chem. Soc.*, **127**, 12492–12493.

17. Kam, N. W. S., O'Connell, M., Wisdom, J. A. and Dai, H. (2005) Carbon nanotubes as multifunctional biological transporters and near-infrared agents for selective cancer cell destruction, *Proc. Natl. Acad. Sci.*, **102**, 11600–11605.
18. Kim, K. Y. (2007) Nanotechnology platforms and physiological challenges for cancer therapeutics, *Nanomed. Nanotech. Biol. Med.*, **3**, 103–110.
19. Kroto, H. W., Heath, J. R., O'Brien, S. C., Curl, R. F. and Smalley, R. E. (1985) C60: buckminsterfullerene, *Nature*, **318**, 162–163.
20. Li, C., Curreli, M., Lin, H., Lei, B., Ishikawa, F. N., Datar, R., Cote, R. J., Thompson, M. E. and Zhou, C. (2005) Complementary detection of prostate-specific antigen using In_2O_3 nanowires and carbon nanotubes, *J. Am. Chem. Soc.*, **127**, 12484–12485.
21. Li, G., Liao, J., Hu, G., Ma, N. and Wu, P. (2005) Study of carbon nanotube modified biosensor for monitoring total cholesterol in blood, *Biosens. Bioelectron.*, **20**, 2140–2144.
22. Liu, Z., Winters, M., Holodniy, M. and Dai, H. (2007) siRNA delivery into human T cells and primary cells with carbon-nanotube transporters, *Angew Chem. Int. Ed. Engl.*, **46**, 2023–2027.
23. Liu, Z., Cai, W., He, L., Nakayama, N., Chen, K., Sun, X., Chen, X. and Dai, H. (2007) *In vivo* biodistribution and highly efficient tumour targeting of carbon nanotubes in mice, *Nature Nanotech.*, **2**, 47–52.
24. McDevitt, M. R., Chattopadhyay, D., Kappell, B. J., Jaggi, J. S., Schiffman, S. R., AntczakC., Njardarson, J. T., Brentjens, R. and Scheinberg, D. A. (2007) Tumor targeting with antibody-functionalized, radiolabeled carbon nanotubes *J. Nucl. Med.*, **48**, 1180–1189
25. Odom, T. W., Huang, J. L., Kim, P. and Lieber, C. M. (1998) Atomic structure and electronic properties of single-wall carbon nanotubes, *Nature*, **391**, 62–64.
26. Okuno, J., Maehashi, K., Kerman, K., Takamura, Y., Matsumoto, K. and Tamiya, E. (2007) Label-free immunosensor for prostate-specific antigen based on single-walled carbon nanotube array-modified microelectrodes *Biosens. Bioelectron.*, **22**, 2377–2381.
27. Pantarotto, D., Briand, J. P., Prato, M. and Bianco, A. (2004) Translocation of bioactive peptides across cell membranes by carbon nanotubes, *Chem. Commun.*, 16–17.
28. Pantarotto, D., Partidos, C. D., Graff, R., Hoebeke, J., Briand, J.-P., Prato, M. and Bianco, A. (2003) Synthesis, structural characterization, and immunological properties of carbon nanotubes functionalized with peptides, *J. Am. Chem. Soc.*, **125**, 6160–6164.
29. Pantarotto, D., Partidos, C. D., Hoebeke, J., Brown, F., Kramer, E., Briand J.-P., Muller, S., Prato, M. and Bianco, A. (2003) Immunization with peptide-functionalized carbon nanotubes enhance virus-specific neutralizing antibody response, *Chem. Biol.*, **10**, 961–966.
30. Pantarotto, D., Singh, D., McCarthy, D., Erhardt, M., Briand, J.-P., Prato, M., Kostarelos, K. and Bianco, A. (2004) Functionalized carbon nanotubes for plasmid DNA Gene delivery, *Angew. Chem. Int. Ed.*, **43**, 5242–5246.

31. Park, D. W., Kim, Y. H., Kim, B. S., So, H. M., Won, K., Lee, J. O., Kong, K. J. and Chang, H. (2006) Detection of tumor markers using single-walled carbon nanotube field effect transistors, *J. Nanosci. Nanotechnol.*, **6**, 3499–3502.
32. Park, K. (2007) Nanotechnology: What it can do for drug delivery, *J. Control Release*, **120**, 1–3.
33. Portney, N. G. and Ozkan, M. (2006) Nano-oncology: Drug delivery, imaging, and sensing, *Anal. Bioanal. Chem.*, **384**, 620–630.
34. Rochette, J-F., Sacher, E., Meunier, M., Luong, J. H. T. (2005) A mediatorless biosensor for putrescine using multiwalled carbon nanotubes, *Anal. Biochem.*, **336**, 305–311.
35. Ruoff, R. S. and Lorents, D. C. (1995) Mechanical and thermal properties of carbon nanotubes, *Carbon*, **33**, 925–930.
36. Singh, R., Pantarotto, D., Lacerda, L., Pastorin, G., Klumpp, C., Prato, M., Bianco, A. and Kostarelos, K. (2006) Tissue biodistribution and blood clearance rates of intravenously administered carbon nanotube radiotracers, *Proc. Natl. Acad. Sci.*, **103**, 3357–3362.
37. Singh, R., Pantarotto, D., McCarthy, D., Chaloin, O., Hoebeke, J., Partidos, C. D., Briand, J.-P., Prato, M., Bianco, A. and Kostarelos, K. (2005) Binding and condensation of plasmid DNA onto functionalized carbon nanotubes: Toward the construction of nanotube-based gene delivery vectors, *J. Am. Chem. Soc.*, **127**, 4388–4396.
38. Takeda, S., Ozaki, H., Hattori, S., Ishii, A., Kida, H. and Mukasa, K. (2007) Detection of influenza virus hemagglutinin with randomly immobilized anti-hemagglutinin antibody on a carbon nanotube sensor, *J. Nanosci. Nanotechnol.*, **7**, 752–756.
39. Tan, X., Li, M., Cai, P., Luo, L. and Zou, X. (2005) An amperometric cholesterol biosensor based on multiwalled carbon nanotubes and organically modified sol–gel/chitosan hybrid composite films, *Anal. Biochem.*, **337**, 111–120.
40. Tang, H., Chen, J., Yao, S., Nie, L., Deng, G. and Kuang, Y. (2004) Amperometric glucose biosensor based on adsorption of glucose oxidase at platinum nanoparticle-modified carbon nanotube electrode, *Anal. Biochem.*, **331**, 89–97.
41. Thomas, C. E., Ehrhardt, A. and Kay, M. A. (2003) Progress and problems with the use of viral vectors for gene therapy, *Nature Rev. Gen.*, **4**, 346–358.
42. Veetil, J. V. and Ye, K. (2007) Development of immunosensors using carbon nanotubes, *Biotechnol. Prog.*, **23**, 517–531.
43. Venkatesan, N., Yoshimitsu, Y., Ito, Y., Shibata, K. and Takada, K. (2005) Liquid filled nanoparticles as a drug delivery tool for protein therapeutics. *Biomaterials*, **26**, 7154–7163.
44. Wang, J., Tangkuaram, T., Loyprasert, S., Vazquez-Alvarez, T., Veerasai, W., Kanatharana, P. and Thavarungkul, P. (2007) Electrocatalytic detection of insulin at RuOx/carbon nanotube-modified carbon electrodes, *Anal. Chim. Acta.*, **581**, 1–6.
45. Wanga, S. G., Wang, R., Sellin, P. J. and Zhang, Q. (2004) DNA biosensors based on self-assembled carbon nanotubes, *Biochem. Biophys. Res. Commun.*, **325**, 1433–1437.

46. Wildoöer, J. W. G., Venema, L. C., Rinzler, A. G., Smalley, R. E.and Dekker, C. D. (1998) Electronic structure of atomically resolved carbon nanotubes, *Nature*, **391**, 59–62.
47. Wu, B. Y., Hou, S. H., Yin, F., Zhao, Z. X., Wang, Y. Y., Wang, X. S. and Chen, Q. (2007) Amperometric glucose biosensor based on multilayer films via layer-by-layer self-assembly of multi-wall carbon nanotubes, gold nanoparticles and glucose oxidase on the Pt electrode, *Biosens Bioelectron.*, **22**, 2854–2860.
48. Wu, W., Wieckowski, S., Pastorin, G., Benicasa, M., Klumpp, C., Briand, J.-P., Gennaro, R., Prato, M. and Bianco, A. (2005) Targeted delivery of Amphotericin B to cell by using functionalised carbon nanotubes, *Angew. Chem. Int. Ed.*, **44**, 6358–6362.
49. Zanello, L. P., Zhao, B., Hu, H. and Haddon, R. C. (2006) Bone cell proliferation on carbon nanotubes, *Nano Lett.*, **6**, 562–567.
50. Zhang, F. F., Wang, X. L., Li, C. X., Li, X. H., Wan, Q., Xian, Y. Z., Jin, L. T. and Yamamoto, K. (2005) Assay for uric acid level in rat striatum by a reagentless biosensor based on functionalized multi-wall carbon nanotubes with tin oxide, *Anal. Bioanal. Chem.*, **382**, 1368–1373.
51. Zhang, M., Mullens, C., Gorski, W. (2005) Insulin oxidation and determination at carbon electrodes, *Anal. Chem.*, **77**, 6396–6401.
52. Zhang, Z., Yang, X., Zhang, Y., Zeng, B., Wang, S., Zhu, T., Roden, R. B. S, Chen, Y. and Yang, R. (2006) Delivery of telomerase reverse transcriptase small interfering RNA in complex with positively charged single-walled carbon nanotubes suppresses tumor growth, *Clin. Cancer Res.*, **12**, 4933–4939.
53. Zheng, M., Jagota, A., Semke, E. D., Diner, B. A., McLean, R. S., Lustig, S. R., Richardson, R. E. and Tassi, N. G. (2003) DNA-assisted dispersion and separation of carbon nanotubes, *Nat. Mater.*, **2**, 338–342.

4

Carbon Nanotubes: Applications for Medical Devices

Robert Streicher

4.1. Introduction

Today, medical implants can be used in almost every organ of the human body. An artificial implant can be defined as a permanent or temporary device to repair or replace a missing biological structure or function. Apart from the biological acceptance of such implants, they ideally should have biomechanical properties that mimic those of the autogeneous tissues they replace without causing any adverse effects. For some applications implants may contain electronics such as artificial pacemaker and cochlear implants; other implants are bioactive, such as drug-eluting stents in the aorta and coronary arteries. The worldwide market for biomaterials is estimated to be in the range of US$ 12 billion. Nevertheless, there is a continuous shift in the requirements and, therefore type and function of the materials needed as well. The therapy for the treatment of diseases or accidents by medical implants is nowadays moving more and more from replacement of the affected tissue, joint or organ to repair and now even towards regeneration, assisting the body's own repair mechanism and requiring ideally only minimally invasive surgery.

Medical technologies and devices are vital and integral components of patient care, and cover a wide range of various therapies and products. Figures have been advanced of over 10 000 different families of medical device types. Given variations in the features of each of these device families, over 400 000 different medical device types could be on the market [Report on the functioning of the Medical Devices Directive 93/42/EC 1993]. It was estimated that the world market for medical devices is around € 160 billion without *in vitro* diagnostic devices [EUCOMED and ADVAMED 2000] and is expected to grow in value by about 9% annually. The European Market, sized at 25.6%, is the second largest market in the world for medical device technology, preceded by US and followed by Japan. Orthopaedic related devices seem to be the biggest part of the business, estimated at about 20% of the total value [The World Medical Markets Fact Book 2006], shown in Figure 1.

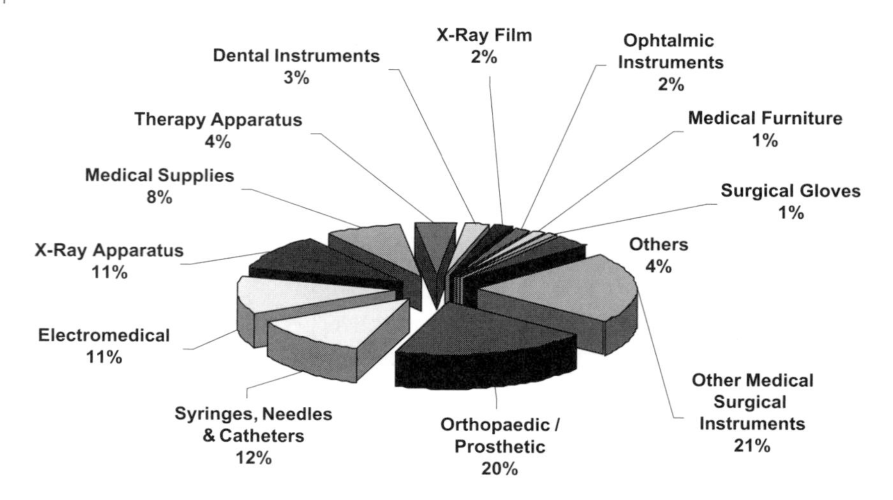

Fig. 1. World-wide market for medical products (2006): prosthetic devices, among them orthopaedic implants, are estimated at 20%.

4.2. Biomaterials and Implants

Materials used for medical implants must fulfill various requirements covering physical, mechanical, biological, toxicological and other aspects,[26] depending on the desired application. Amongst the required properties, biocompatibility plays a key role in the performance[108] and has been defined as the ability of a material to perform with an appropriate host response in a specific medical application.[132] In other words: an implant should be compatible in order to stay in the human body over a defined time, interact properly with living tissue and cause no inflammatory, toxic, immunogenic, or mutagenic reaction. The biological evaluation of medical devices is governed by a set of standards developed by the International Organisation for Standardization (ISO) and known as ISO Standard 10993 or, in the US, as US Food and Drug Administration (FDA) blue book memorandum #G95-12, which is a modification of ISO 10993. In addition, the environment should not cause degradation or corrosion of the material that would result in loss of its physical and mechanical properties, if not desired as with biodegradable biomaterials.

The development and production of implantable medical devices have already started even before World War II. A common feature of the materials used, which were derived from engineering applications, was their biological inertness. The availability of biomaterials was an important step to be able to address the medical needs of patients suffering from severe trauma or chronic disabling diseases such as osteoarthritis, cardiovascular dysfunction, etc. In the 70s research and development moved from bio-inert to bio-conductive to bio-active and resorbable biomaterials, which for the first time were specifically engineered for medical applications. By the mid-80s it was possible to produce intra-corporeal medical devices and components thereof, provoking a biological tissue response that could elicit controlled actions and reactions within the body. By the end of the 80s, the first bio-active materials appeared in a variety of musculoskeletal devices, amongst them glasses, ceramics, and composites, as well as bio-resorbable polymers. Originally a technology associated with mechanical engineering, technological developments in the medical device sector with e.g. tailored materials have shifted boundaries. With tissue engineering entering the traditional sectors of healthcare, the traditionally clear borderlines between pharmaceuticals and medical devices are blurring, and information and telecommunication technologies have added new dimensions, introducing at the same time enormous benefits and complexities.

4.2.1. *Orthopaedic Implants*

Because of its importance, specific attention is paid to the orthopaedic segment, which deals with the therapy and correction of deformities, injuries and other disorders of the musculoskeletal system. Numerous artificial devices are available to stabilize bone fractures, spinal deformities or fractures, or to replace joints, for example hip, knee, finger and shoulder. Several million people are affected by orthopaedic diseases and trauma, as depicted in Figure 2,

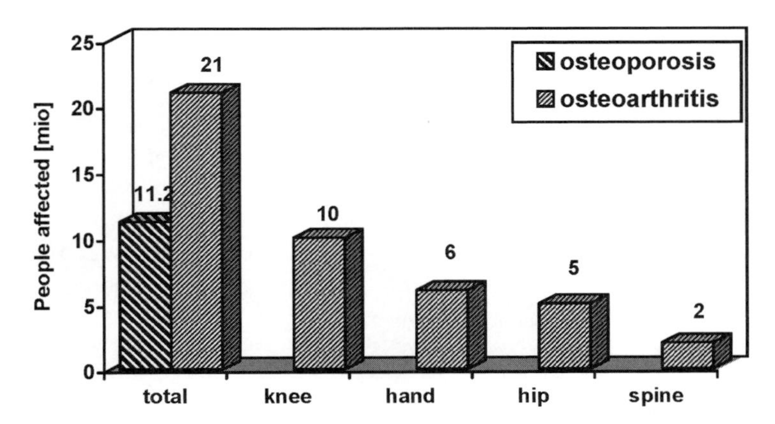

Fig. 2. US market data (2005) for orthopaedic related problems. 7.8% of the 11.2% of osteoporosis are asymptomatic, the remaining 3.4% need treatment for related bone fractures at various locations.

and the market size for orthopaedic devices is in the range of US$30 billion worldwide [The World-Wide Orthopaedic Market 2004–2005, Technological and Surgical Advancements Shaping the Orthopaedic Industry 2006], with more than a million devices per year being implanted.

The annual growth rate for implants for restoration of the locomotive apparatus is estimated to be 10% (20–25% for spinal implants) due to the increasing incidence of musculoskeletal diseases like osteoporosis, accident-related fractures and osteoarthritis. Other key factors driving growth are the ageing population with their related problems as well as injuries related to increased mobility and the increased patient demand due to the changing lifestyles.

The development of orthopaedic implants has evolved over several decades, leading to better devices with regard to a reduced complication-rate and increased longevity. Although the devices are very successful, there are still several unresolved issues and it is expected that with the higher demand from patients and also smaller or bone conserving implants, new devices and new materials are necessary to avoid the current issues of implant fracture, allergic reaction, missing radio-opacity, wear, corrosion, degradation, etc. About 10% of all implant interventions comprise revisions for failed implant-tissue integration, with increasing incidence [The World Medical Markets Fact Book 2006]. The subsequent revision surgery is demanding and takes longer, and recovery takes weeks or months and is less successful than the primary procedure. Also the cost of revision surgery exceeds the costs of a primary intervention by far. The reasons for implant failure are multi-factorial, but the main cause is aseptic loosening of the components (75% of all revisions).

For the production of implants and devices for trauma, sports medicine, spinal and other musculoskeletal disorders, materials from almost all available groups are used: metals for load-bearing components like plates and nails for fracture fixation, rods for spinal fixation and stems etc. for total joint implants. Structural ceramics find their application in articulation components of total joints, while non-biostable ceramics based on calcium-phosphates are used as bio-conductive coatings or fillers. Polymers can be used for articulation components of total hip joints or spinal implants, bone cement for fixation of joint prostheses or for stabilizing vertebrae, as structural components, as temporary resorbable fixation devices or scaffolds, and in many other musculoskeletal applications for soft and hard tissue repair.

4.2.1.1. *Current Issues with Orthopaedic Implants*

Although considerable progress has been made since their introduction, and the success rate of replacements for the hip (THA; Figure 3(a)) or knee joint

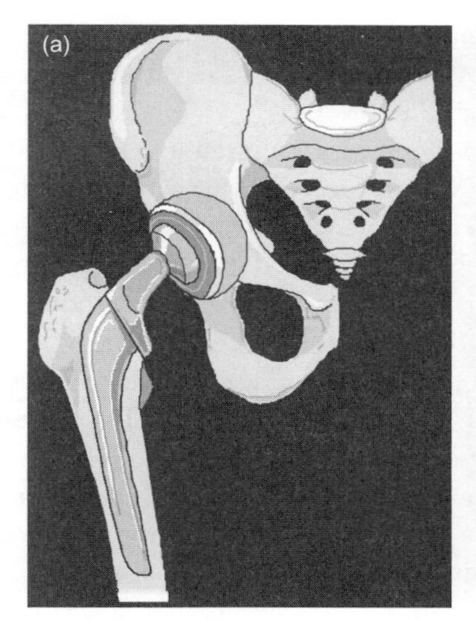

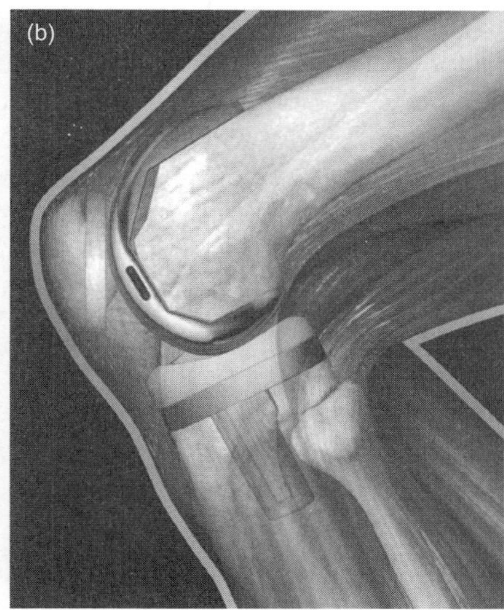

Fig. 3. (a) Schematic of a hip joint replacement, consisting of stem and cup component (b) Total knee joint, consisting of tibial, femoral and patella replacement.

(TKA; Figure 3(b)) is high, a revision rate of 10% is expected within the first 10 years of their use.

Aseptic loosening is the main cause of failure and is usually a late complication often caused by an inflammatory response to wear particles from the articulation of joint replacements.[115] Insufficient or non-lasting osseo-integration as well as stress-shielding, a biomechanical phenomenon occurring with relatively stiff implants, are other reasons for the failure of the implant-tissue interface resulting in aseptic loosening of the implant and the potential of peri-prosthetic fractures.

Almost all of the materials used as bearings in total joint arthroplasty are the same as or incremental modifications of materials in long-term clinical use, some of them for more than 40 years. Ultra High Molecular Weight Polyethylene (UHMWPE) has been used since 1962 for THA cups, tibial components of TKA or articulating components of artificial spinal disks. Metal-on-metal and ceramic-on-ceramic articulations are in use for specific designs and mainly for young and demanding THA patients. All of the biomaterials used for joint replacement have several disadvantages. Polyethylene (PE) is not highly wear resistant and is susceptible to creep, degradation and fracture. Due to its poor strength and high creep it has design limitations such as the necessary thickness to reduce the risk of deformation or fracture. Metals release metal ions into the peri-prosthetic tissue, which can cause systemic effects like hypersensitivity and also cancer. Alumina ceramic is brittle and stress sensitive and has therefore a certain fracture risk and design limitations.

The natural stress distribution in bone is significantly altered after fixation of a traumatic fracture or total arthroplasty. Any implanted device will carry some of the load normally transferred through the bone, causing a change of stress distribution in it. If bone is not loaded appropriately it will remodel according to Wolff's law,[122,134] resulting in either bone mass loss through resorption (atrophy) or building bone in the regions of higher stress (hypertrophy). Metal or ceramic implants or components thereof, with a Young's modulus of 100–300 GPa, are considerably stiffer than bone with a Young's modulus of 8–24 GPa, shown in Figure 4. Bone loss is associated with stress shielding from a mismatch between the modulus of the bone and the implant[73] and can lead to implant loosening. This does not seem to be an issue for trauma applications, where the implant is only acting as a temporary stabilization device, but it is an issue for reconstructive long-term implantable devices like artificial joint replacement.

The unmet needs and unsolved problems for orthopaedic implants can be summarised as follows:

(1) Safe and effective biocompatible materials for the intended purpose.
(2) Enabling small and thin but stable and biomechanically sound components allowing conservation of as much healthy and functional tissue as possible.

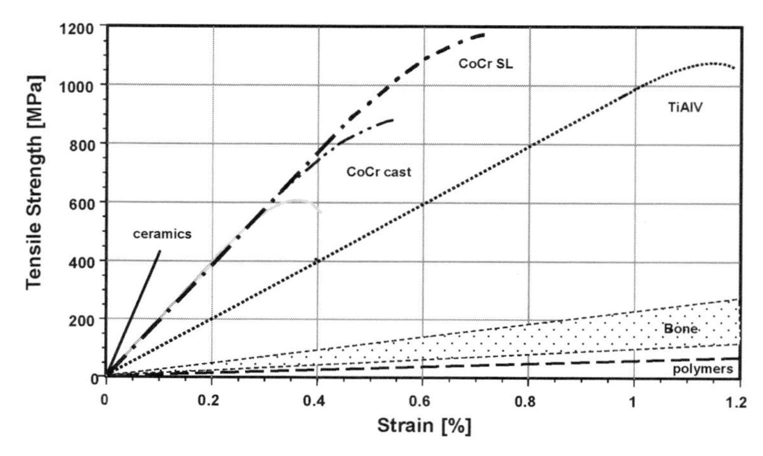

Fig. 4. Mechanical properties and Young's modulus mismatch of existing biomaterials (ceramics, metals, polymers) used for orthopaedic devices compared to bone.

(3) Fast and long-lasting fixation between the implant and the host tissue to allow appropriate stress transfer.
(4) Wear-resistant bearing surfaces with low friction.

Any improvements addressing those issues will have a major impact on better performance and longer lifetime of total joint arthroplasty, and also on the economic outcome of such an operation.

4.3. Nanomaterials in Medicine

Nanotechnology provides opportunities not only to improve materials and medical devices, but also to create new smart devices and technologies like for example intelligent drug delivery systems.[95,123] Several review papers deal with the various applications of nanomaterials and nanostructures for human health care.[76,102,104] Artificial nanostructures and biomaterials have been suggested and investigated for numerous medical applications to enhance healing or replace tissues and organs, etc. in the form of coatings, films, structured surfaces, scaffolds or composites.[77]

Being hollow nanofibers, nanotubes offer interesting features. Tubes are lighter than full structures and offer higher mechanical strength. As drug carriers, tubes have the potential to store and release much higher quantities than surfaces. Therefore, out of all nanofibers, nanotubes are of the highest interest for several applications in medicine. Nanotubes, especially short ones, can be manufactured from several materials, potentially filled with for example therapeutic substances and also closed at one or both ends ("capped"). Their surfaces can also be functionalised by attaching various functional groups increasing their hydrophilicity and reactivity.[48]

4.4. Carbon Nanotubes

Carbon nanotubes (CNTs) are part of the fullerene family, which was discovered in 1985 by researchers at the University of Sussex and Rice.[67] CNTs were first observed and reported by Iijima.[55] While fullerenes — a family of carbon allotropes — are spherical in shape, nanotubes are cylindrical with at least one end typically capped with a hemisphere of fullerene structure. They have a high aspect ratio, being only a few nanometers in diameter but up to a millimetre long.

There are two main forms of nanotubes: single-walled (SWNTs) and multi-walled (MWNTs). A SWNT is a one-atom thick sheet of graphite rolled up into a seamless cylinder, while an MWNT structure consists of multiple layers of concentric SWNTs. Depending on the ways of rolling up the hexagonal graphene sheets, CNTs are either of the armchair, zigzag or chiral variety, leading to different properties. The properties of nanotubes — unusually high strength, unique electrical properties, chemically inertness and extremely high thermal conductivity — make them useful not only in electronic and mechanical applications but also for medical applications. The elastic modulus of SWNTs is about 1 TPa. CNTs have a low density for a solid of 1.3–1.4 g/cm^3, and the highest specific strength of all known materials. In addition to this CNTs are chemically inert and thermally stable up to 750°C in air.[20] Since their discovery, several methods have been used to produce single and multi-walled nanotubes. These include plasma arc discharge, pulsed lasers, chemical vapor deposition[133] and are described as well in another chapter of this book.

Although CNTs show excellent properties, they have some characteristics which make them difficult to use. Their high molecular weight and the strong intertube forces keep CNTs, especially SWNTs, together in agglomerates, making their manipulation difficult. Functionalisation, i.e. the attachment of appropriate chemical functional groups onto their conjugated carbon scaffold, offers several advantages. The derivative tubes exhibit improved properties with respect to solubility and ease of dispersion, manipulation and processability. Functionalisation is also a necessity for achieving an appropriate interfacial bonding to any matrix system for nano-composites to enable good stress transfer between the CNTs and the matrix,[62] and also provides a site to attach therapeutic agents or chemical groups for transport, targeting and therapeutic use in medicine.

4.4.1. *Biomedical Applications*

With their carbon composition, high aspect ratio, mechanical, electrical and physical properties, there has been growing interest in using CNTs for medical and biomedical applications like orthopaedic and dental implants and neural probes. Since some years, the number of articles related to their use for biomedical applications has approximately doubled annually.[46] The biologic activity and bio-kinetics of any biomaterial depend on many parameters, and the requirements and testing methods are well established and regulated for the classic materials. However, due to the different physiochemical properties associated with their size, CNTs can potentially provoke a response in the human body that is different from, and not directly predicted by, the constituent chemicals and compounds. For example, even a traditionally inert bulk compound, such as gold, may become bioactive in the nanometer range.[43] Too many parameters and variables at this level determine the interaction of CNTs with cells and their supra-structures for any intended application. Additionally, CNTs themselves are currently not reproducibly manufactured and processed (size, shape, distribution, defects, charge, purity, crystallinity, solubility, bio-persistence, functionalisation, etc.), and, therefore not well characterised. Apart from variations between different manufacturers, even batch to batch variability is high. Because of this enhanced complexity the biological qualification of CNTs is not clear at this stage, even for cell cultures or small animal species. Over the last years a series of investigations has been published and the knowledge about the toxicity and biocompatibility of CNTs in their various forms is increasing.[35,36] Two chapters in this book deal with these issues, and therefore this key property for their use in medical applications is not discussed in this chapter.

CNTs are under investigation for potential use in several biomedical applications, given their capacity to interact with macromolecules such as proteins and oligosaccharides, and their nanoscale dimensions which are similar to those of elementary biomolecules such as enzymes, proteins and DNA.[11,17] They hold promise for applications in medicine, drug and gene delivery areas. Carbon nanotube arrays can play a key role in artificial cochlea development.[47] There is a series of other potential applications of CNTs and composites thereof for medical applications like for example MWNT nano-composites as biomimetic sensors, actuators or artificial muscles.[71] The ability to change the surface chemistry and properties by functionalisation of the sidewalls of CNTs has facilitated their manipulation and also paved the way to other biomedical applications such as vascular stents, or platforms for neuron growth and regeneration. CNTs are also being investigated as delivery vehicles for anticancer drugs directly into cancer cells[4,77] or as nano-bombs to selectively destroy cancer cells.[60,112] A futuristic approach to curing diseases was proposed by Bhargava[12] who suggested the use of nano-robots made of CNTs, because of their strength and chemical inertness, for example protecting the immune system by identifying and attacking bacteria and viruses.

4.4.2. *Interaction with Biologic Structures*

When biomaterials get into contact with cells, their surface structure, chemistry and charge define the biological response. Because all biologic processes, including traumatic or pathologic events, are governed and influenced by cell interactions at the nano-level,[16,38,125] carbon nano-structures with their remarkable properties have been proposed for several applications in direct contact with living tissue to allow specific cell attachment or rejection. One of the most promising medical applications for CNTs and carbon nanofibres (CNFs) apart from drug delivery, etc. is tissue engineering. This involves modifying and using biomaterials for building scaffolds to regenerate functional tissue by influencing the organisation, differentiation and growth of appropriate cells for body functions that have been lost or impaired as a result of disease or accident. CNTs and CNFs offer compelling properties for such applications, but their surface chemistry usually needs to be adapted to achieve the desired biologic reaction. A recent publication by Harrison and Atala provides an excellent overview on the use of carbon nanotubes for tissue engineering.[46]

Tissue engineering is of particular interest in the healing of any form of traumatical, disease related (e.g. avascular necrosis, tumor) or artificially created wound and gap, because the organism has only limited ability to bridge such distances. Without scaffolding, the defect would heal within the gap to form extended scar tissue that would jeopardise structure, form and function of the repaired tissue. Scaffolds can be 2-dimensional (e.g. skin care) or 3-dimensional structures (e.g. bone replacements). 2D structures are usually achieved by surface modifications, like addition of CNTs to substrates or as nano-filler in composites. The surfaces of various materials can be structured with nano-patterns by incorporating or applying CNTs for improved tissue-implant interaction. Just as important as the topography is the chemistry of the surface. 3D structures, which allow transmission of all forms of load, are normally built up by regular or irregular fibrous structures with interconnecting pores that allow the cells to penetrate and fill the scaffold. It has been shown that MWNTs can be shaped into 3D architectures and are ideal for cell seeding and *in vitro* cell modelling, leading to the design of promising new tissue-engineered products for biological applications. In 2004, for example, the growth of a mouse fibroblast cell line on a 3D sieve architecture network based on an array of interconnected MWNTs by exerting chemically induced capillary forces upon the nanotubes was reported.[22] The network supported cell attachment and growth and possessed enough structural integrity and stability to retain its shape in vivo, with adequate mechanical strength to support the developing tissue. Such a nano-material could serve as a biocompatible matrix to restore, maintain or reinforce damaged or weakened tissues or for use where MWNTs can act as drug delivery device.

There is a defined trend for medical devices to move from tissue substitution by artificial replacements to guiding and accelerating the healing process by means of tissue engineering, which can yield enormous personal and economic benefit. Due to their unique properties, CNTs can play a major role in the development of appropriate structures and surfaces. They have also been proposed recently for use in tissue engineered dental applications, delivered as an alginate nano-composite gel at the site.[61] Nevertheless, it needs to be kept in mind that tissue engineering is not usually intended as permanent structure in the organism, but only to assist the healing process. Ideally, these scaffolds are replaced by living structures over time and the artificial structure is either metabolized or excreted. Because of their covalently bound carbon atoms, CNTs and CNFs are very stable and can only be excreted or, less favorable, be encapsulated by macrophages. It must be well understood what happens with CNTs after they have served their purpose either as carrier, sensor, scaffold, reinforcement, etc. For i.v. administered, DTPA-functionalised and radio-labeled MWNTs and SWNTs Singh *et al.* found excretion via the urinary passage within a half-time of three hours.[111] Deng *et al.* used 14C-labelled MWNTs for i.v. injection in mice and found accumulation in the liver and hepatic macrophage encapsulation.[28] Although the toxicity was found to be relatively low, it also shows how crucial an appropriate functionalisation is. Other authors[18] also tracked i.v. administered SWNTs in rabbits and found that after circulating for more than one hour they were excreted via the liver. No acute toxic effect for low doses was reported, but trace amounts were still found in the kidney.

4.4.2.1. *Tissue Engineering for Neural Applications*

The central nervous system consists, simplified, of neurons and neuroglia. Neurons, the "nerve cells", have a core area (soma) with a diameter of up to 100 mm, equipped with extensions (dendrites) serving as input device for signals from other neurons. Nerve signal output is guided by the axon which can be as long as 100 cm and ends in branches (terminals) connecting to dendrites of other neurons. Axons and dendrites are subsumed as neurites. The neuroglia consists of several types of glial cells. There are approximately 10 times more glial cells than neurons. The neuroglia not only gives physical structure to the brain but handles the entire infrastructure of the neuronal metabolism. The synaptic cleft where neurotransmitters are emitted from an axon terminal to the dendrite of the receiving neuron has a width of approximately 20 nm. Other "fine" structures of neurons, dendrites and cell walls are nanosized as well. And, most interestingly, the neuroglia does not totally fill the space between neurons, blood vessels and other structures but everywhere leaves a space of 20 nm. This means that all metabolic and nervous functions are transmitted through nano-sized gaps. If these key processes are analyzed or influenced, nano-sized probes are needed, if possible electrically conducting to be able to interact with nervous signals.

To allow continuous monitoring, diagnosis, and treatment of neural tissue, implantable probes are required. Neuronal devices need to interact electrically to be able to stimulate neuron cell axonal outgrowth for repairing damage areas of the central and peripheral nervous system and also encourage and maintain cell on-growth or in-growth.[32] As clinical experience with neural probes made of silicon has shown, they can become encapsulated with non-conductive, and therefore undesirable, glial scar tissue.

Carbon nanofibers (CNFs) and carbon nanotubes (CNTs) offer unique mechanical, optical, and most importantly excellent conductivity properties, which may be beneficial in the design of more effective neural prostheses. Carbon nanofibers have a less regular arrangement of helical carbon than carbon nanotubes, and inferior mechanical properties to the latter.[62] Both nanomaterials have high strength to weight ratios, are chemically inert or can be appropriately functionalised, and have excellent electrical conductivity. They also have comparable surface structures and surface chemistries, both of which are considered suitable for neural applications. It has been shown that any foreign body response decreases with increasing conductivity of a material,[106] and electrical stimulation seems to be beneficial for potential applications as central and peripheral neural biomaterials for nerve regeneration and functions.[66] Most ways of using CNFs and CNTs and composites made thereof for candidate materials to interact with neuronal tissue have been suggested only recently.

Webster *et al.*[125] assumed that the structures of neural cells are naturally accustomed to interact with nanostructures. They proposed CNTs as a means to create brain probes and implants for studying and treating neurological damage and disorders. For a series of experiments they developed and used medical grade polycarbonate urethane (PCU) reinforced with various amounts of non-functionalised CNF, and monitored the attachment of rat nerve cells. One type of cells was neurons, the other astrocytes, a glial cell type. Neural devices need preferential attachment of neurons, whereas glial cells stand for neurally inactive encapsulation that limits functionality. They conducted several experiments to determine the electrical and mechanical properties of composites with CNF content varying from 0 to 100 wt%. More importantly, they demonstrated an increase in neural cell functions and a simultaneous decrease of glial-scar forming astrocytes with increasing carbon fibre content of the composite. They concluded that surface factors were leading to the selective adhesion of cells, and therefore decreasing the potential for device encapsulation by fibrous-scar tissue formation. They speculate that the surfaces of the composites with high CNF content have more nanometer irregularities, which are on the same scale as laminin, a cruciform-shaped protein found on the surface of the brain. The neurons can recognize parts of the protein and latch onto it, which helps neurons sprout neuritis, and therefore reduces scar tissue formation. Their study provided first evidence of the potential of CNFs and composites for interacting with neural cells to enable the development of successful neural probes and implants. In an attempt to explain the observed phenomenon Khang *et al.*, from the same group, investigated in a subsequent study the surface characteristics of the PCU composites, with CNF content varying from 0 to 30 wt.%, for osteoblast reaction.[64] They demonstrated that the investigated surfaces yield higher roughness with higher effective surface area, and also increase the surface energy by enhancing the CNF concentration, leading to enhanced osteoblast adhesion.

Another study by the Purdue group looked into the effect of the diameter and the surface energy of the CNF PCU composites.[84] First, they seeded rat astrocytes onto the substrates and measured adhesion, proliferation and long-term function. The results demonstrated that the glial scar tissue forming cells preferentially adhered and proliferated on large diameter and low surface energy (25–50 mJ/m2) CNFs. Then they repeated the experiments with a PCU composite with 0–25 wt% high surface energy (125–140 mJ/m^2) CNFs, and found less adhesion of astrocytes with an increasing amount of CNFs. They concluded that positive interactions with neurons[83] and, at the same time, limited astrocyte functions will reduce the potential for gliatic scar tissue formation and therefore improve the efficacy of a neuronal implant. Liu and Webster[76] reported recently that researchers have started to impregnate CNFs with stem cells to reverse rat brain damage induced by stroke, and concluded that their *in-vivo* results demonstrated the potential to treat such massive neurological disorders.

Several other research groups also investigated the potential of CNTs for neural tissue engineering. CNTs are not biodegradable, and as such they could be used as implants where long-term extra-cellular molecular cues for neurite outgrowth are necessary, such as in regeneration after spinal cord or brain injury. Mattson *et al.* concluded that the conductivity of CNTs coated with bioactive molecules promotes neuron growth.[83] Hu *et al.* chemically attached

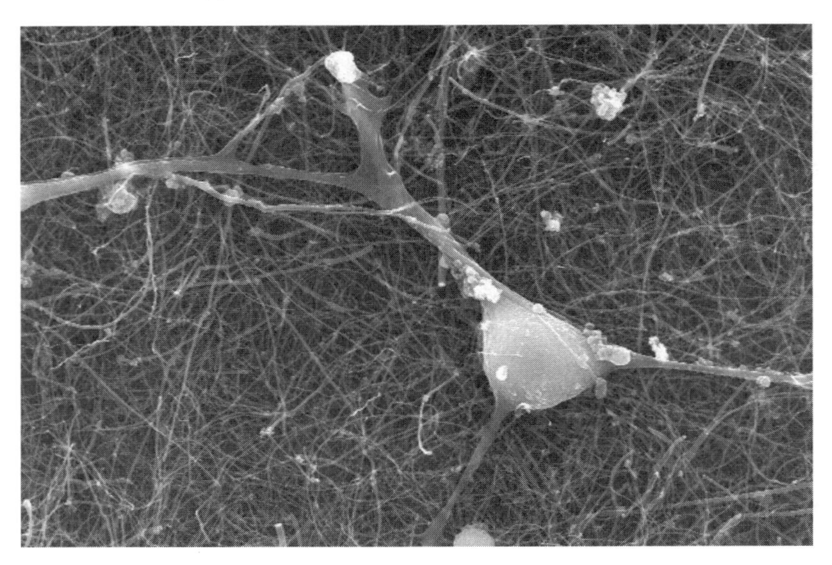

Fig. 5. A hippocampal neuron in culture growth on MWNTs. Modified from Hu *et al.*[56] (Courtesy of V. Parpura, University of California, Riverside, US.)

various functional groups to MWNT and used them as substrates for cultured neurons to investigate potential applications as neural prostheses;[54] the result is shown in Figure 5.

According to their results CNTs could act as an extra-cellular scaffold to guide neurite outgrowth and regulate neurite branching, controlled by the charge of the functional groups, leading to the re-establishment of the intricate connections between neurons and forming synapses, which has potential applications e.g. for regeneration after spinal cord or brain injury. Further work by the Riverside group[92] was performed with functionalised water-soluble single-walled carbon nanotube (SWNT) graft copolymers, prepared by chemical functionalisation with poly-m-aminobenzene sulphuric acid and polyethylene glycol, for modulating the outgrowth of neuronal processes. They observed in a culturing medium that those SWNTs were able to increase the length of various neuronal processes. In 2003, the same group[44] issued a patent describing a method of studying neurite outgrowth at the nanometer scale by employing functionalised CNTs as substrate. First the carbon nanotube is functionalised with a neuronal growth promoting agent, then immobilized on a glass cover slip and placed in a culture dish with a culture medium. This is followed by the seeding of neuronal cells into the culture dish. Due to the functionalisation of the CNTs, the neurites grow along the fibers and, thus, can be directed.

The invention ultimately targets implants promoting nerve regeneration. This concept seems to be feasible because CNTs support long-term cell survival, and, therefore seem to be an ideal substrate for this purpose.

Liopo *et al.* used a model neuronal cell with unmodified and functionalised SWNT films with 4-benzoic acid or 4-tert-butylphenyl groups.[75] They demonstrated that neuronal attachment and growth was supported, and that SWNTs did not interfere with ongoing neuronal function. The primary peripheral rat neurons could be electrically coupled to SWNTs, and showed robust voltage-activated currents when stimulated through this conductive film. The data suggest that SWNTs could be used for applications involving electrically excitable tissues such as muscles and nerves.

Another group of researchers[72] used CNTs and CNF as carriers for neural stem cells implanted into rat brains which were damaged in a controlled manner. Re-established electrical activity in the damaged brain areas showed that the implanted stem cells, after up to eight weeks, developed into viable, functioning neurons. The authors expressed their hope that this could be used in future for positively influencing pathological conditions such as Alzheimer's disease. Taking advantage of the electrical conductivity and the small dimensions, MWNT were used to wire a nano-scale integrated circuit, designed for neural electrophysiological imaging and for prosthetic devices in the nervous system.[27] The MWNT were directly grown on the chip by CVD and insulated with PMMA polymerized *in situ*.

Gabay *et al.* presented a novel approach for patterning cultured neural networks by fabricating a nano-patterned array of high density CNT islands on a silicon dioxide chip using a template method based on photolithography, micro-contact printing and CNT chemical vapor deposition.[41,42] These sites served as electrically viable, neuronal networks, consisting of ganglion-like clusters of neurons and, at the same time, as electrodes to record neuronal signals. Although the effect of CNTs on neurons is unclear, they concluded that their work provided an innovative engineering approach to form complex interconnected neuronal networks with pre-designed geometry via self-assembly of neurons. Sorkin *et al.* from the same research group assumed that the electrical conductivity of the CNTs used as substrate is beneficial and concluded that carbon nanotubes can be used as structural support for engineered tissue.[114]

In their work, Xie *et al.* were able to demonstrate that functional groups on CNTs can act as anchoring points to enhance the adhesion and extension of neurons and neurites and promote neurite growth.[137] Dubin *et al.* obtained "hair like" electrically conductive and highly porous CNFs by coagulation

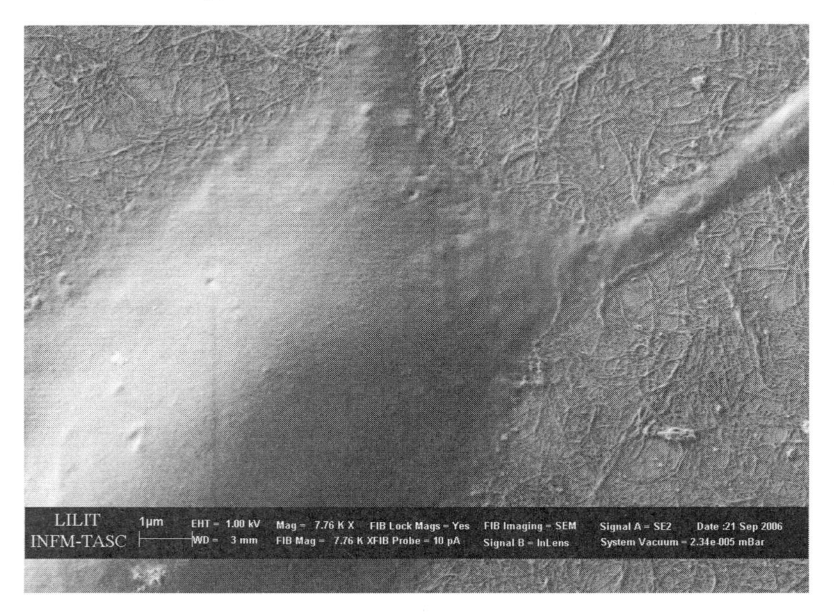

Fig. 6. Hippocampal neuron grown in culture for 8 days on purified SWNTs. (Courtesy of S. Campidelli, M. Prato, L. Ballerini, University of Trieste, Italy.)

spinning and used them as substrates for rat dermal fibroblasts.[29] These fiber bundles showed no toxicity to the attached and growing cell lines. The authors concluded that these fibres could be used for the development of electrically conductive interfaces with neuronal tissue. Lovat *et al.*[78] observed cell adhesion, dendrite elongation and increase in the efficacy of neural signal transmission on a CNT grid and attribute this to the specific properties of CNT materials, such as the high electrical conductivity. The neurons deposited on the CNT grid, depicted in Figure 6, show a significant increase in the frequency of spontaneous postsynaptic currents compared to neurons grown on glass with no effect. They postulate that this may be due to the electrical conductivity of the substrate and thus that CNTs can be used as structural supports for engineered tissue. Another research group[117] reported on a nanocomposite consisting of blends of poly-lactic-acid and CNTs, which could be used to expose cells to electrical stimulation.

The unique combination of nano-sized surface, structural, chemical and more importantly electrical properties make CNTs and CNFs, especially if functionalised, ideal substrates for neuron proliferation, for analyzing electric signals or stimulating them, and for promoting and guiding neurite outgrowth.

Placed in arrays on electronic circuits, dots from CNTs and CNFs are the locations where single neurons attach and start to reach out with their axons to connect to other neurons' dendrites. Although the positive effects of carbon nano-structures have mainly been demonstrated in animal cell cultures and need to be researched and explored, the implications of potential neuronal processors for better understanding the function and dysfunction of biologic systems are gigantic. In future nano-probes might be widely used for the synaptic cleft and other intercellular spaces, an area where today only a glimpse of what seems feasible can be seen. Several applications can be thought of, e.g. for individuals who have lost limbs or are paraplegic. It might be possible to e.g. move their limbs again by direct brain activation and control of exo-prostheses (artificial limbs), connected via CNT modified interface to sensor/transmitting biological tissue etc. Another possibility is to reverse brain injuries after trauma by stimulating cell regeneration via the local application of CNTs, etc.

4.4.2.2. *Tissue Engineering for Orthopaedic Applications*

There is also a defined similarity of bone tissue and nanostructures, with bone being a nano-composite of inorganic hydroxyapatite (HA) crystallites (2–5 × 20–50 nm) in an organic collagen matrix (0.1 − 8 × 300 nm) in the nanometer range. Because tissue development is controlled by events at the cellular and molecular level, the surfaces for osteointegration need to be able to influence osteoprogenitor population activity and function. Several recent review papers deal with the similarity of bone and carbon nano-structures and report the use of CNFs and CNTs.[8,19,77,126] Since some years, research is looking into developing bio-mimetic tissue surfaces for prostheses; surfaces that have been engineered at the nano-scale to mimic or interact with soft or hard tissue, in effect behaving as a living surface. With the possibilities that nano-technology offers, many current problems in developing long-lasting bio-compatible orthopaedic devices might be solved in the future. Several applications for nanostructures and nano-materials are already being pursued, and first results are promising.[116] The use of CNFs and CNTs has been proposed and investigated either as a stand-alone product, especially if functionalised, as scaffold or as a filler/reinforcement for polymers or other composite materials for various orthopaedic applications, mostly focusing on hard tissue (bone) repair.[8,19,77,126]

Price and colleagues compared polymer casts of consolidated conventionally sized and nano-sized carbon fibers, highlighting the importance of the nanostructure. This study demonstrated a clearly better osteoblast adhesion to the nano-phase fibers.[100] In a subsequent study they produced composites with the same fibers using a polymer matrix of poly-carbonate-urethane (PCU). Such composites have already attracted attention for orthopaedic applications because of their tailorable electrical and mechanical properties through the integration of carbon nanofibers. In their experiments they demonstrated again the advantage of 10 and 25 wt% CNFs for osteoblast adhesion.[99] Webster and others[30,125] demonstrated in various other papers a selective adhesion of osteoblasts on carbon nanofibers (CNFs), and on composites containing such, which makes them potential candidates for prosthetic devices. They used carbon nanofibers and CNTs in a matrix of PCU to add a nanostructure to the surface, mimicking nano-features of living tissue, and tested them with cell lines *in vitro*. They found that selective adhesion was more pronounced with increased content of functionalised CNT/CNF: osteoblasts showed increased viability[30] and were preferred over fibroblasts.[100] In another study with the same nano-composite it was shown that also fibronectin adsorption increased with an increasing amount of MWNTs, a consequence of the increased surface roughness and surface energy of the nano-composite.[63] Osteoblasts align along CNFs with a pyrolytic outerlayer, i.e. reduced surface energy, and depose calcium phosphate mineral, which is seen as a sign of their viability.[64] All of these phenomena would decrease fibrous tissue formation around an implant and are desirable for optimized osseo-integration.

Shi *et al.* reported recently another variation of an injectable, *in situ* crosslinkable nano-composite based on poly-propylene-fumarate and low concentrations of pristine and functionalised SWNTs, in an effort to design new orthopaedic biomaterials with mechanical properties suitable for load-bearing bone engineering applications.[108,109] They concluded that the functionalised SWNTs were better dispersed and also enhanced the mechanical properties of the composite, even with addition of only 0.1 wt% SWNTs, making it a promising material for the intended purpose.

Looking for a novel bone substitute, the use of chemically functionalised SWNTs for tissue growth scaffolds for the treatment of bone fractures has been suggested.[143] This approach combines the well known osseo-conductive properties of hydroxyapatite (HA) *in vitro* with those of CNTs, adding tensile flexibility and strength to the brittle bioceramic bone substitute. For this purpose, SWNTs were chemically functionalised with several chemical groups to create negative surface charges that attract calcium ions from the surrounding electrolyte. Some of these attached groups led to nucleation, self-assembly and orientation of the HA crystals, allowing control of their alignment, while other groups improved the biocompatibility of the CNTs to increase their

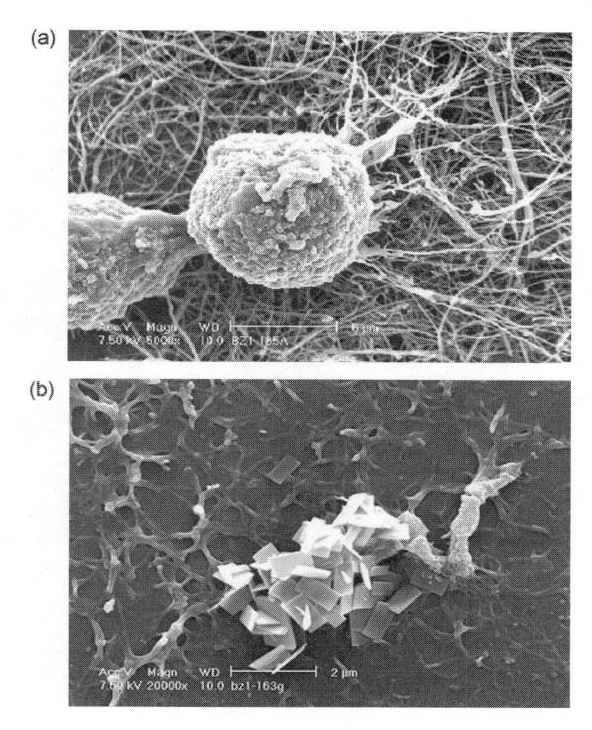

Fig. 7. (a) Dividing osteoblast on MWNTs (b) Mineralized matrix found in SWNT cultures. (Courtesy of L. Zanello, University of California, Riverside, US.)

solubility in water. This work can lead to artificial bone with improved flexibility and strength, new types of bone grafts and, even more importantly, to a potential local treatment of osteoporosis. This could be done by simply injecting a solution of functionalised nanotubes into bone defects or fractures to encourage new tissue to grow and heal defects. Zanello *et al.* showed *in vitro* that osteoblast cells grow and proliferate best on scaffolds of SWNTs and MWNTs, shown in Figures 7(a), (b), when the functionalisation leads to a neutral surface charge.[138]

Another approach to increase the durability and biological acceptance of surfaces for applications as orthopaedic implants is the use of coatings consisting of CNTs and HA. Balani *et al.* attempted to reinforce HA plasma spray coatings with 4% MWNTs and reported that the fracture toughness of the nano-composite is improved and no adverse cellular reactions were detected in *in vitro* tests.[7] White *et al.* give an overview on HA-CNT composites for biomedical applications.[131] An example is shown in Figure 8.

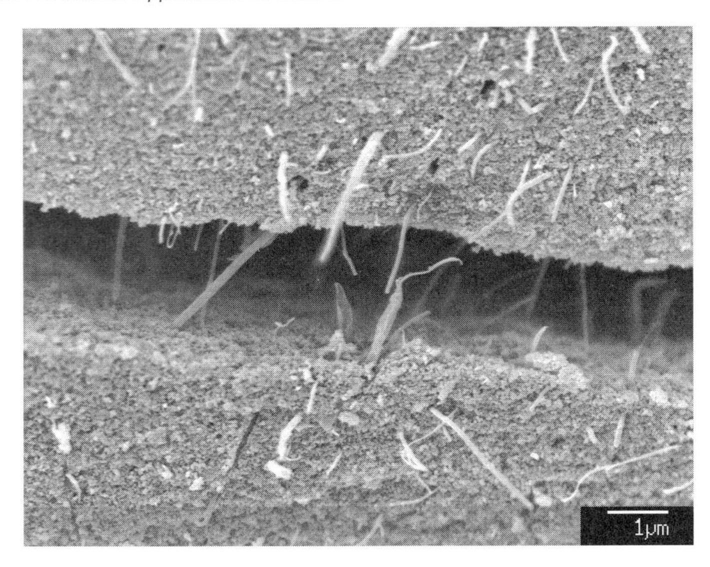

Fig. 8. SEM micrograph of HA-CNT composite (Courtesy of S.M. Best, University of Cambridge, UK.)

Another nano-composite with CNTs for biomedical application can be produced by preparing a MWNT/poly-L-lactide (PLLA) composite, demonstrating that the interaction between the polymer and MWNT occurs mainly through the hydrophobic C-CH_3 functional groups.[141] The conductivity of the composite increases as the MWNT content is increased and the growth of fibroblast cells is inhibited. The properties of this biomaterial allow for potential applications in orthopaedics and other fields of medicine requiring a biodegradable conductive material with selective cell interaction capacity. Nevertheless, the crucial question about the fate and the biologic reaction of the released CNTs emerging from any resorbed matrix material needs to be addressed.

Electrospinning can be employed to create CNT-containing fibres, for example SWNT-reinforced silk nanofibers[5] and adding only 0.8 wt% of MWNTs to a chitosan biopolymer matrix doubled its tensile strength.[124] MacDonald *et al.* integrated *in vitro* up to 4% SWNTs into collagen, to enhance strength and also to achieve an electrically conductive composite with cells.[79,80] The authors expect that such collagen-CNT composite matrices may have utility as scaffolds in tissue engineering. Correa-Duarte *et al.*[22] used chemically induced capillary forces to interconnect MWNTs to a regular 3D-network and successfully grew mouse fibroblast cells in this structure.

Firkowska *et al.* combined lithography with a layer-by-layer self-assembly process to achieve highly ordered 3D matrices based on intercrossed MWNTs for potential use in tissue engineering.[37]

4.4.3. *Structural Components for Orthopaedics*

A thin and flexible, but strong acetabular cup prosthesis that could mimic the normal transmission of weight-bearing forces to the supporting bone should significantly reduce bone loss and provide long-term stability. First attempts with a composite construction of a flexible horseshoe-shaped artificial acetabular cup, shown in Figure 9, have already been made, and Jones *et al.*[58] reported about the use of a CF reinforced poly-butylene-terephthalate composite cup replacement with good results in 50 octogenarian patients after five years in clinical application.

The design used has a lot of room for improvements. If a monoblock composite cup based on such a concept could be developed which encourages

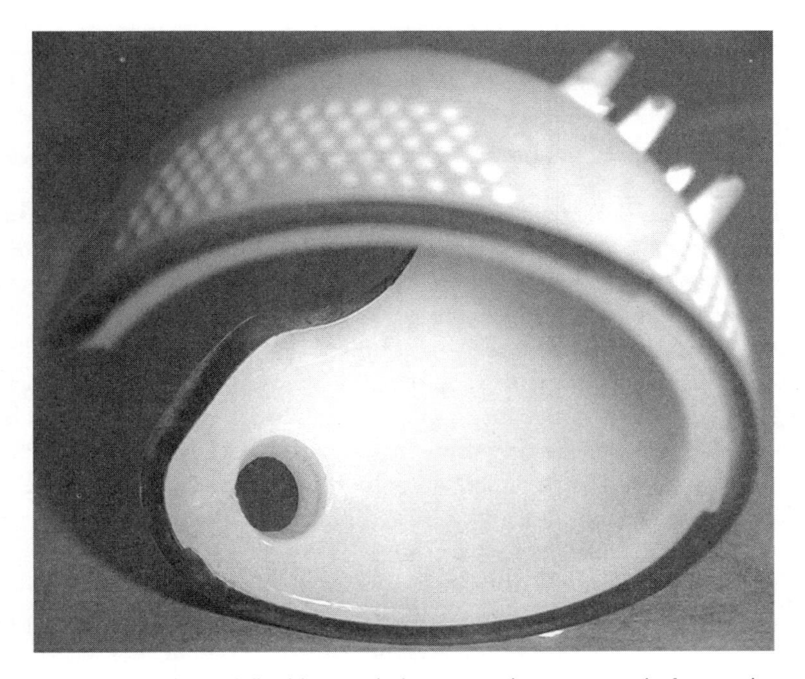

Fig. 9. Horseshoe formed flexible acetabular cup replacement made from polymer-CF composite with articulating layer of UHMWPE and HA surface coating.

long lasting bone ongrowth on the outer surface and is at the same time wear resistant on the inner surface, this would be an outstanding improvement enabling a new, bone-conserving and biomechanically sound cup prosthesis. Similar possibilities are available taking into consideration the current trend for segmental or partial joint replacement, e.g. resurfacing of the hip joint, which is currently achieved by using large diameter metal-metal articulations for young and active patients. There is an ongoing interest to conserve as much natural tissue as possible to allow for bone stock preservation and for any revision operation, as the life expectancy of a younger patient population is higher than that of the implants. The interest in increasingly smaller implants also pushes the limits for the existing materials, and there is a need for new materials and technology to satisfy the criteria of such designs. Several applications for spine treatment (e.g. radio-opaque cages filled with osseo-active or conductive substances, artificial disc), trauma treatment (e.g. radio-opaque plates, nails, non-corroding screws, etc.), and the treatment of degenerative diseases also need improvement. Nanotechnology and CNTs have the potential to achieve the goal of such developments, especially when using all properties and not just the mechanical ones, and could have a major impact on new designs and implants.

4.4.3.1. *Carbon Fibers and Composites for Orthopaedic Applications*

The history of carbon fibers (CF) for intra-articular orthopaedic applications is discouraging. CFs have a bad reputation because they can penetrate the cell membrane and lead to apoptosis. Blazewicz gives an excellent overview on the history and status of various carbon materials which have been applied for the treatment of soft and hard tissue injuries.[13] Recently carbon nanofibers, 155 nm in diameter, and nanographite particles <1 micron in diameter, have been shown to have a significant adverse response in cell cultures for osteoblasts and fibroblasts, depending only very little on particle concentration, but causing severe apoptosis for the nanofibers at high concentration.[20] There are also publications drawing an analogy between such high-aspect-ratio nanostructures and asbestos fibers.[49,82]

Composites comprise a matrix of a material or structure and a reinforcing agent of another material or structure, and have applications in sophisticated engineering and highly demanding aerospace components. Composite materials have the potential to be adapted to suit particular biological environments; for example their properties may be tailored for better mechanical

Table 1. Summary of polymers and fibers used in various combinations for structural medical applications

Matrix Materials	Fiber Materials
Poly-sulphone	Carbon/graphite
Poly-ether-ether-ketone	Kevlar
Epoxy	CNTs?
Triazine	
Ultra-high Mw Poly-ethylene	
Poly-butylene-terephtalate	

compatibility and strength for an individual purpose and application. Composites have been manufactured from various materials, summarized in Table 1, and used for various implants in medicine and orthopaedics.

Some of those composites have only been used in limited clinical trials. In orthopaedics the use of carbon fibers and/or composites made thereof started around the end of the 1970s / beginning of the 1980s with artificial ligaments made of woven CF,[40,56,85] which abraded and/or fractured causing degeneration of the knee joint.[23] Composites of CF in an epoxy matrix have also been used as plates for internal fracture stabilization. Although biomechanical and preferable to metal plates because of their radio-opacity, the tissue carbon staining observed in the peri-prosthetic tissue[53] was a cause of concern.

At the end of the 1970s an orthopaedic company introduced components of artificial joint implants made from a composite consisting of a matrix of ultra-high molecular weight poly-ethylene (UHMWPE), the material of choice for artificial joints since 1962, and 10% carbon fibers.[2,45] The claims with these products were that the inclusion of carbon fiber reduces the level of creep of the composite, and also increases the resistance to wear. Figure 10(a) shows the lack of bonding between the UHMWPE matrix and the non-treated CFs of this composite, explaining its poor fracture resistance.

As a result, fiber detachment was likely to occur and cause surface pitting, an effect already observed by Peterson.[96] The detached hard carbon fibers consequently scratched the surface of the articulating metallic component, and exacerbated the wear issue. A further problem was the increased contact pressure as a consequence of stiffening the UHMWPE by the addition of the fibers. The material therefore experienced a higher stress than conventional UHMWPE for a similar design. This stress could exceed the strength of the composite and thus affect the failure mechanism. Furthermore, the effects of the low creep resistance of UHMWPE enhanced the de-bonding of the fibers in an already poorly structured material.

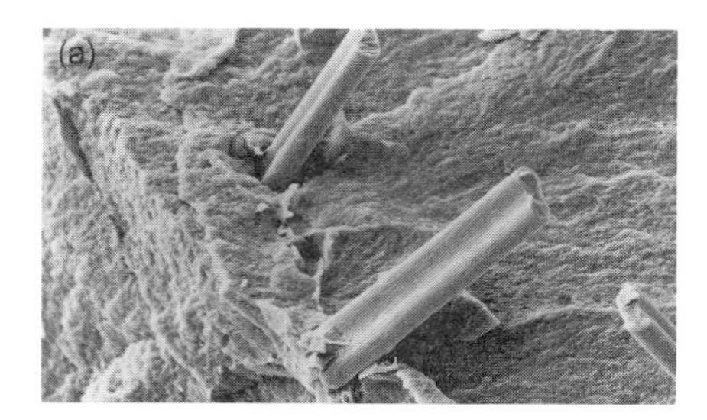

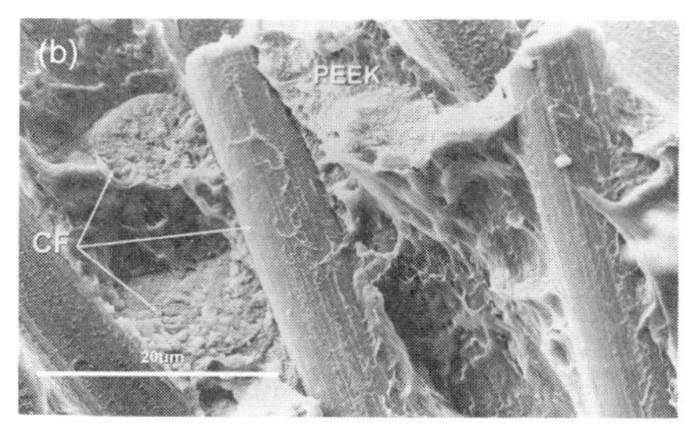

Fig. 10. (a) Fracture surface of a UHMWPE-CF composite showing insufficient adhesion between matrix and fibers. (b) Fracture surface of a PEEK-CF composite: the fibers show partial intimate contact with the matrix which increases mechanical strength of the composite.

Several papers deal with the negative experience with this CF reinforced composite.[14,24,101,136] Wright *et al.*[135] reported fatigue crack propagation test results which showed that the propagation rates for CF-UHMWPE were 8 times higher than those of conventional UHMWPE. Pryor *et al.* reported about 2 clinical cases with failed hip replacements, and they observed staining of the peri-prosthetic tissue despite low wear of the composite.[101] The histological analysis revealed an atypical foreign body reaction with histiocytes within the collagen stroma, and they also found within all of the cells tiny intra-cytoplasmic fragments of carbon. The reaction to the wear particles of the composite was different to that of the peri-prosthetic tissue from failed hip implants with non-reinforced UHMWPE, and they concluded that this was the reason for the early failure of the total hip joint. This clinically unsuccessful composite serves as an example of what needs to be optimized for any new generation of fiber reinforced polymer composite.

Further developments with polymer CF composites started already in 1980s for stems (poly-sulfone CF), which fractured during the clinical investigation. At the end of the 80s composites of Triazin-CF and epoxy-CF were used for monoblock cups, without much success due to the lack of appropriate designs and the fixation of the implants to the bone. The interest has focused on potential bearing surfaces with CF reinforced polymers seen as possible alternatives to metal alloy components.[25,33] Thermoplastic polymers with their strong intermolecular bonds resist moisture damage, conferring good biocompatibility and durability.[57,88,129] Materials investigated were CF reinforced poly-ether-ether-ketone (PEEK),[74,98,142] poly-sulphone (PSO)[81] and amorphous carbon-CF composites.[9,13,89,113]

PEEK is a high strength thermoplastic polymer used for high-tech applications, with excellent characteristics that include chemical resistance, and has been qualified as a biocompatible material in its bulk form, available in a medical grade formulation. It is used for spinal cages, having the advantage of being strong, stable and radio-opaque. A composite for application as a modular acetabular insert seated in a shell of titanium alloy for a total hip replacement has been developed based on PEEK as matrix. Integrated into the matrix were 30 wt% milled pitch based carbon fibers of 200 microns length and 11 microns diameter. A fracture surface is depicted in Figure 10(b) showing partial bonding between CF and the matrix, despite no treatment of either. This composite has been utilized in industrial applications,[68] and has excellent tribological properties as well as biological characteristics.[87] The composite was undergoing extensive biological tests according to the ISO 10'993 standard for the bulk materials, including extracts of the composite in polar and non-polar solvents. Grinding the composite resulted in <15 micron particles, which also underwent biological testing. The results on biocompatibility demonstrated that all aspects of the composite, ranging through bulk products, leachable extracts and wear debris, are well tolerated.

Its high strength (Figure 11) makes the composite attractive for structural and wear resistant applications in orthopaedics and high fracture energies for pure PEEK and CF-PEEK have been determined.[70] The high fracture energy for crack propagation is reflected in the toughness of the composite, with further benefit resulting from the fact that the PEEK matrix is also able to exhibit local plastic deformations during impact loading and prevent cracking.

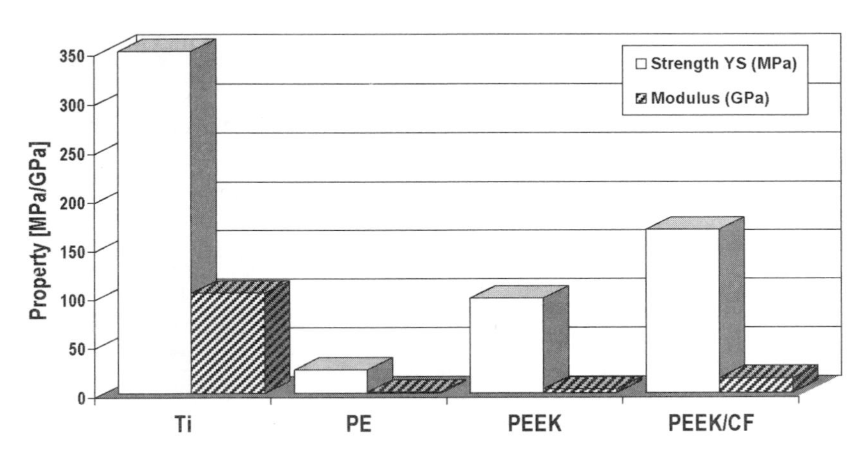

Fig. 11. Comparison of selected mechanical properties of orthopaedic biomaterials: The PEEK/CF composite presents high strength and low stiffness which is attractive for structural biomedical components.

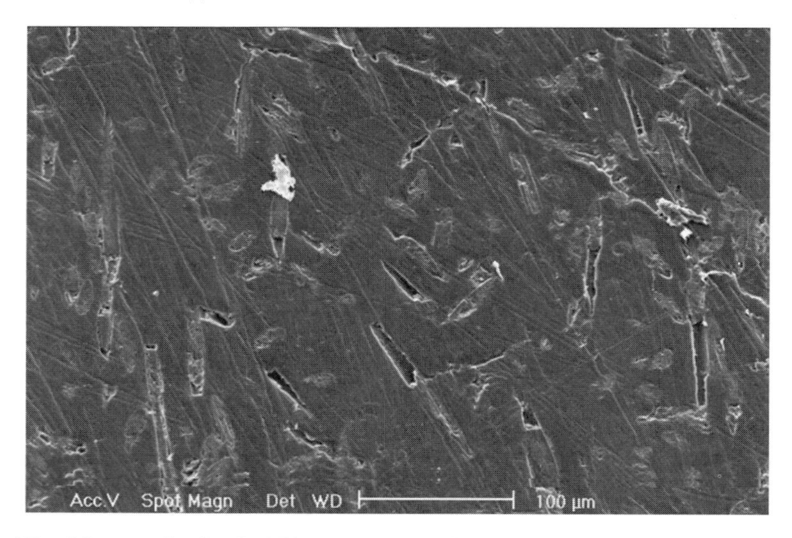

Fig. 12. Micrograph of a CF-PEEK composite surface after 5 million cycles in a hip joint simulator. Scratches and fiber pull-out.

The creep resistance of the PEEK composite is also high under the expected loading conditions, and therefore not an issue.

Hip joint simulator test results gave an average wear rate of 0.4 mm^3 per million cycles, which is a factor of at least 10 less than for the conventional wear couple of metal or alumina articulating against UHMWPE[59] and even lower than for metal-on-metal articulation (1.6 mm^3 per million cycles). Nevertheless, the wear surface of the composite exhibited roughness with some fiber pull-out, as shown in Figure 12, and leaves room for improvement.

Another potential application for the same polymer-CF composite has been proposed for the femoral component of a knee joint prosthesis articulating against soft polymers.[105] In 2001 a limited clinical evaluation with the composite as a cup insert for total hip joint prostheses was started, and the early experience of that composite showed good clinical results at one year.[94] Given the observed partial integration of the CFs in the matrix and resultant fiber pull-out, several improvements for high performance composites can be envisioned.

4.4.3.1.1. Polymer/CNT Nano-composites

To enhance the strength and mechanical resistance of long-term implants, the reinforcement of high strength polymers with carbon nanotubes (CNTs) seems to be a promising route. CNTs are highly anisotropic and due to their high strength, high aspect-ratio and excellent thermal and electrical conductivity, they are being used instead of carbon fibres for making composites. They have generated a great deal of interest particularly as fillers in polymer matrix composite materials. CNTs are thought to improve the mechanical, electrical and thermal properties of these composites. However, several difficulties related firstly to the dispersion of nanotubes and their compatibility with the matrix, and secondarily to the nature of the bonding between them, need to be resolved prior to practical applications. To exploit the unique properties of CNTs in a polymer composite, fundamental challenges need to be addressed that have proven to be a complex issue with a number of competing factors[90,128]:

1. Production and purification of the CNTs: single or multiwalled, diameter, aspect ratio, wall defects, orientation, etc.
2. Functionalisation of the CNTs: physical and/or chemical, temperature, etc. and subsequent modification of their structure.
3. CNTs in the composite matrix: dispersion, concentration, alignment, bonding, etc.
4. Composite processing method; matrix material; interaction CNT-matrix, crystallinity of the matrix, thermal history, etc.

A strong interfacial bonding is necessary to allow an appropriate stress transfer between CNTs and the matrix of nano-composites.[69] Three main load transfer mechanisms control the main stress transfer: surface topography and micromechanical interlock, chemical bonding, and van der Waals interaction. To achieve a bonding between the reinforcement and the polymer matrix two approaches have been followed. The first is non-covalent attachment of molecules, leading to only weak interfacial van der Waals forces. The second is covalent attachment of functional groups to the walls of the nanotubes. The non-covalent approach has the advantage that the perfect structure of the nanotubes is not altered, and their overall mechanical properties remain the same. Covalent attachment of functional groups to the surface of nanotubes can be achieved by a multitude of chemical and/or physical methods. This might introduce defects on the walls of the perfect structure of nanotubes and consequently lower the strength of the reinforcing component. It has also been shown to have an effect on the stiffness of CNT,[91] which needs to be taken into account. Nevertheless, it is assumed that the shear strength of a polymer-CNT interface with non-covalent bonding can be increased over an order of magnitude by introducing $<1\%$ of chemical bonds between the nanotube and matrix. Another aspect is the effect of any filler or reinforcement on the structure of the matrix. The incorporation of any substance during the processing of a thermoplastic semi-crystalline polymer can have an effect on its crystallinity, and therefore also affect the properties of the CNT composite. The trade-off between the strength of the interface, strength of the nanotube and strength of the composite itself must be well balanced.[102]

Much research has focused on the development of CNT-polymer composites that have a high potential as lightweight high strength fiber-reinforced materials for non medical applications, demonstrating that, apart from the polymer, the CNTs and any applied functionalisation, there is also a trade-off between the amount of CNFs necessary to achieve a significant change in the properties of the composite and its processability.[65] Several methods have been applied to achieve a strong interface between the CNTs and various polymer matrixes. Single walled (SWNT) and multi walled (MWNT) CNTs have been utilized for reinforcing thermoset polymers and were also investigated for thermoplastic polymers, including PEEK.[119,120] For example it has been shown that functionalisation of nanotubes with amino acids by heating oxidized CNTs with an excess of triethylenetetramine leads to covalent bonding of the nanotube surface with an epoxy resin. Ami Eitan *et al.* have demonstrated functionalisation of SWNTs and MWNTs by using a mixture of sulfuric acid and nitric acid to form carboxylic acid groups on the surface that are formed along the nanotube walls and the ends, which seems to be ideal for optimal load transfer within the composite.[3] They state that it is also possible to further react the epoxide functional group to enable even better interaction between the polymer matrix chains and the CNT surface of the composite.[128]

Only a few abstracts so far mention the potential application of CNT composites for structural orthopaedic applications. In an attempt to improve the mechanical properties of UHMWPE, a well accepted hydrophobic and non-polar articulation material for artificial hips and knees, by adding functionalised CNTs to it, reinforced composites have been manufactured by researchers from Sweden[31] and Montpellier.[1,6] The group from Montpellier used UHMWPE with non surface modified SWNTs as a reference, which resulted in non covalent attachment between nanotubes and the polymer. Although by this approach the CNT structure is not altered, the mechanical properties of the composite would only be affected minimally and the low attraction forces between matrix and nanotubes are disadvantageous,[50] as the clinical history of the same polymer has shown.[101] They also covalently linked nanotubes and polymeric matrix both with chemically modified SWNTs and PE.[6] This was achieved by oxidizing the nanotubes forming carboxylic acid groups along their walls and ends, and thermally oxidizing the UHMWPE, forming oxygen-containing functional groups and using a di-amine to link the oxidized PE and SWNTs. The di-functional molecule can react with the carboxylic acid groups attached on both SWNTs and PE to form a strong covalent bond. The resulting composites were manufactured only as film or as fibers. Mechanical testing of the resulting composite showed improvements over the reference composite. Another interesting aspect of using CNTs to reinforce UHMWPE is the observation that MWNTs can act as radical scavengers and antioxidants,[34] especially in view of the fact that the medical grade of UHMWPE is only available without any antioxidant and tends to degrade in the body environment.

PEEK, a semi-crystalline high strength polymer, has been investigated more than most other polymers for nano-composite applications. A composite with the addition of 10% MWNTs is commercially available for technical applications. Shaffer *et al.* reported on the mechanical properties of a vapour-grown 150 nm diameter carbon nanofiber (CNF) reinforced PEEK composite using extrusion to produce master-batches. The final samples were prepared by using an injection molding process. A linear increase in stiffness and yield strength with loading fractions up to 15 wt% was observed, while the PEEK matrix ductility was maintained up to 10 wt% CNF filling. No effect of the filler on matrix crystallinity was detected. Werner *et al.* investigated the tribological properties of this composite.[130] They tested specimens in a set-up using an ISO 7148-2 ball-on-prism test system. An exponential wear rate reduction for composites with 5–10% CNF loading was observed, depending weakly on the filler content. Their theory is that CNFs act as solid lubricant, thereby reducing the wear rate.

With the intention of producing structural orthopaedic implants, Babaa *et al.* also used PEEK as a matrix material and reported the fabrication of UHMWPE-CNT and PEEK-CNT composites.[6] To obtain a covalent bond between the components, they chemically functionalised the CNTs by using di-functional molecules for a better bonding between CNT and the polymer. With this approach the properties of the composite were improved. At a recent meeting Bantignie reported about the results with such functionalised MWNTs and sulfonated PEEK using a direct attachment reaction bonding to the sulphur group of PEEK, and demonstrated using spectroscopic methods that a covalent bonding between SWNTs and the PEEK matrix could be obtained.[10] Despite the cumbersome process and the question of its commercial feasibility, functionalised CNTs have been shown to allow an intimate bonding with functionalised PEEK, a clinically accepted and used material.

This concept offers an excellent opportunity for load-bearing applications of thin and high strength implants with exceptional wear properties, which could also be engineered on their interface with the host tissue to be fully integrated.

For another application, the fixation of endo-prosthesis to bone, the addition of 0.005–5.0% single or multiple walled CNTs for enhancing the properties of poly-methyl-meth-acrylate based bone cement has recently been proposed and a patent filed.[97] The addition of the CNTs has a stronger effect in avoiding crack propagation in this amorphous polymer than in a semi-crystalline polymer, and would address one of the major faults of the brittle bone cement: fracture and fatigue fracture. Positive side effects are that such a reinforced cement would shrink less and exhibit a lower exothermic temperature rise, and that the hollow nanotubes would allow for incorporation of pharmaceutical compositions like antibiotics, anti-inflammatory substances, chemotherapeutical agents, etc. Potential issues may be an increase in the viscosity of such a CNT-enhanced bone cement during application, a potential phase separation between the cement and the CNTs while filling the gaps between implant and bone, and a change in its setting behavior. Another potential effect that needs to be looked at is the interaction between the CNTs and the metallic implant with regard to corrosion, fretting and abrasion, especially with a higher CNT content.

4.4.3.1.2. Alumina/CNT Nano-composites

Alumina-CNT composites are of interest for several potential medical applications, especially joint implants. Increasing the low toughness and bending strength of alumina by enhancing the resistance to crack propagation would allow for more design flexibility, less fracture risk and better wear resistance. This would be beneficial for implant components made of alumina, which at present have certain limitations due to the fracture risk e.g. of thin articular cups or inserts, small heads for hip implants or thin femoral components for knee implants.

The literature also reports on enhancing the properties of metals and ceramics by producing metal-CNT and ceramic-CNT composites, and several methods have been patented.[51,52,119–121,139] As a brittle material, alumina is strong in compression but is susceptible to fracture, represented by two values, flexural strength and fracture toughness K_{1c}. The alumina quality used for artificial joint bearings has a typical flexural strength of 500 MPa and a K_{1c} of $4\,MPa^*m^{1/2}$. To enhance its resistance against crack initiation and propagation, either fibers or toughening constituents, which can absorb or deviate parts of the strain energy induced by a stress or crack, are used. CNTs could

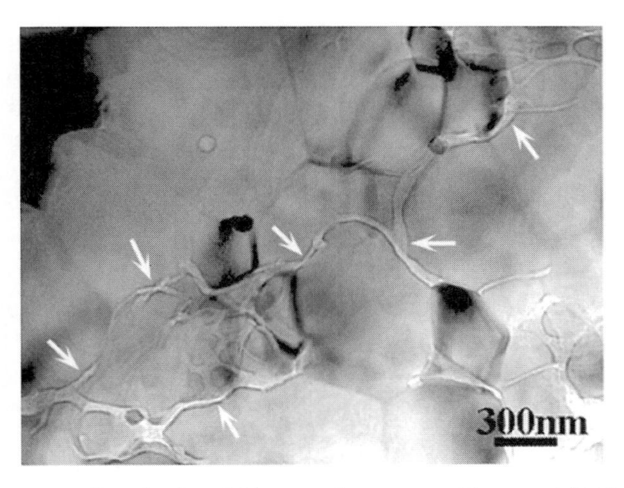

Fig. 13. Nanocrystalline alumina-CNT composite structure (Courtesy A.K. Mukherjee, University of California, Davis, US.)

have such an effect at the nano-scale. For producing CNT-alumina composites various approaches have been proposed,[15,86,140] but little data is reported on the effect of adding CNTs. In Figure 13, the structure of such a composite is depicted, also illustrating the problem of achieving a fully dense and homogenous structural material.

Despite this, Zhan *et al.*[140] reported successful fabrication a dense nano-crystalline alumina-SWNT nano-composites at sintering temperatures as low as 1150°C by spark-plasma sintering, and achieving a fracture toughness of 9.7 MPa*$m^{1/2}$, nearly three times that of pure nano-crystalline alumina. Established alumina ceramics are sintered at temperatures above 1400°C in an oxidizing atmosphere, which would burn the imbedded CNTs. Thus, the composites have to be sintered at much lower temperatures of about 1000°C. It is doubtful whether oxide ceramics meeting today's strict standards can be produced like this. One solution to a lower sintering temperature is to allow more glassy phases between the ceramic grains, e.g. by using lower purity alumina, but this reduces overall quality. A more promising approach to higher strength ceramics might be the use of nano-structured ingredients.[118]

4.5. Summary and Conclusions

The need for new technologies and implants in medicine for the treatment of severe accidents and diseases is increasing. This is due not only to an increasing population age with the related degeneration process and diseases, but also to changing expectations and lifestyle with a growing demand for fast solutions to health-related issues. New ideas, concepts, materials and devices are required to enable less tissue damaging and more tissue regenerating, innovative medical devices implanted in a minimal invasive and guided way to allow for accurate local placement and fast recovery. Especially for implants, which are supposed to substitute a patient's degenerated or absent function in the long-term, several challenges must be addressed. Such implants would benefit dramatically from the availability of a multi-functional material that can fulfill both biomechanical and biological requirements while at the same time offering considerable design freedom. As the need for new approaches to primary interventions increases, the increasing number of more demanding revision operations also needs to be addressed at the same time.

The biomaterials available today are reaching their limits, despite their successful use for treating various medical conditions. They must meet several requirements covering physical, mechanical, biological, toxicological and other aspects, depending on final application. Carbon is chemically inert, and carbon nanotubes not only exhibit superior mechanical, chemical and electrical properties, but also seem to be biocompatible especially when purified and/or functionalised. Functionalisation positively influences the interface properties, which facilitates the handling and administration of CNT suspensions and allows for intimate contact between CNTs and matrices in composites. Various functional groups (polar, non-polar, acidic, basic, etc.) can be attached to such a reactive surface, thus offering a high potential for unique applications in wide areas of medicine, and the number of publications and patents for such applications is increasing dramatically. Also, the encapsulation of other materials within carbon nanotubes would open up extended applications for therapeutic use in medicine.

Despite all enthusiasm, there are some characteristics and restrictions to be kept in mind. The field of nano-materials for medical applications is still in its infancy and numerous questions including manufacturing, regulatory and economic issues must be addressed, apart from the biological and technical ones yet to be defined. Preliminary results increasingly confirm the high potential of CNTs for medical applications either as a structure, a coating, a scaffold or reinforcement for composites, although most of these are at laboratory scale and *in vitro* testing. Much work still has to be done to address safety issues, as CNTs are not biodegradable materials and therefore need to be cleared from the organism, e.g. after injecting CNTs for diagnostic purposes or when a composite matrix resorbs.

A number of fundamental issues also need to be resolved. CNTs can be produced by various methods and in various forms. Apart from the various forms, and therefore potentially different reactions of different types of cells and living tissues to such variations, it seems that they cannot yet be reproducibly manufactured or processed and batch-to-batch variations cannot be excluded. For all applications, but especially medical ones, nanotubes need to be purified to a high level without changing the properties and function of the CNTs. This challenge in attaining a homogenous, reproducible and well characterised product also makes the definitive biological qualification of CNTs complex and difficult, so that the question of their potential toxicity is still open. Individual treatment of the carbon nanotubes for direct or indirect functionalisation adds to the complexity of determining and predicting their biological/toxicological behavior even in cell cultures or small animal species. If CNTs are to be applied in structural composites the crucial interaction with the matrix must be resolved, mainly by functionalisation of one or both components. As functionalisation also changes the properties of the individual components and the composite overall, the potential improvement in fiber dispersion, integration and load transfer must be balanced against any undesirable modifications.

It also seems from the current literature that is not yet clear how nanostructures and above CNTs will behave in the long term in a living environment, and how they may change their properties with time. The versatility of CNTs for changing and modifying almost every single property may become an issue in a permanently changing environment like the human body, and there is a clear need for continuous, more comprehensive monitoring of the potential effects of newly designed and fabricated devices in *in vivo* long term situations.

Nevertheless, studies on the potential applications of carbon nanotubes highlight several future opportunities in cardiovascular, neurological, dental, orthopaedic and other medical disciplines. Several potential applications have been proposed and researched, such as new tools for nano-diagnostics including biosensors and medical imaging, regenerative medicine, and drug delivery and release systems for therapeutic purposes.

Apart from those promising approaches to biological applications, CNTs have also been shown to be attractive for selective surface interaction with various host tissues and bio-matrices for directing cell growth, given their nano-scale dimensions that are similar to those of elementary bio-molecules, and their capacity to interact with biological macromolecules. Preliminary results have demonstrated the increasingly preferred interaction of CNT surfaces with cells and tissues, while reducing the affinity for scar tissue developing progenitor cells, for the next generation of engineered biomaterials and medical devices. Therefore, CNTs have already been proposed as surfaces and scaffolds for various applications of tissue regeneration for neurological, orthopaedic and other medical applications. They have also been suggested and researched for soft tissue applications like ligaments, tendons, muscles, cartilage, meniscus, etc. First attempts have also been made to apply CNTs in combination with other biomaterials in composites for structural implants, replacing existing solutions which have certain restrictions with regard to design flexibility, biomechanical response, wear resistance and tissue interaction.

Nano-materials and especially carbon nanotubes have an enormous potential for changing therapeutic measures and methods for trauma and pathological situations in an aging population. New materials, technologies and designs will allow in future for an armatorium to treat any patient on the principle of implementing autologeous tissue and processes in rebuilding affected tissues and structures. Even if partial or complete substitution of the natural structure is unavoidable for saving the patients life or regaining life quality, CNTs can play an important role by enabling thinner components with greater design flexibility, enhanced tailor-made properties and at the same time enhanced functional tissue integration. However, there is a significant need for interdisciplinary teamwork and exchange at multiple levels to successfully address the current issues, before being able to fully explore the enormous potential of carbon nanotubes for medical devices.

Bibliography

1. Abountanos, V., Babaa, M.-R., Holzinger, M., Fiorito, S., Poncharal, P., Steinmetz, J. and Zahab, A. (2005) Fabrication of Composite of UHMWPE and Functionalised Nanotubes, Scientific Report CNRS Montpellier 2002–2005, 28.
2. Ainsworth, R., Farling, G. and Bardos, D. (1977) An improved bearing material for joint replacement prostheses: Carbon fiber reinforced UHMW polyethylene, *Trans. 3rd Soc. Biomat. Meeting*, 11.
3. Ami Eitan, K. J., Dukes, D., Andrews, R. and Schadler, L. S. (2003) Surface modification of multiwalled carbon nanotubes: Toward the tailoring of the interface in polymer composites, *Chem. Mater.*, **15**, 3198–3201.
4. Apler, J. (2006) Nanotubes poised to help cancer patients. *Joe Apler Monthly Feature* — January 2006, http://nano.cancer.gov/news_center/monthly_feature_2006_jan.asp

5. Ayutsede, J., Gandhi, M., Sukigara, S., Ye, H., Hsu, C. and Gogotsi, Y. (2006) Carbon nanotube reinforced bombyx mori silk nanofibers by the electrospinning, process, *Biomacromolecules*, **7**, 208–214.
6. Babaa, R., Bantignie, J. L., Michel, T., Fiorito, S., Poncharal, P. and Zahab, A. (2006) Carbon nanotubes reinforced PE and PEEK: Fabrication and characterization, *Abstr. 20th IWPE Molecular Nanostructures*, PTue5.
7. Balani, K., Anderson, R., Laha, T., Andara, M., Tercero, J., Crumpler, E. and Agarwal A. (2007) Plasma-sprayed carbon nanotube reinforced hydroxyapatite coatings and their interaction with human osteoblasts *in vitro*, *Biomaterials*, **28**, 618–624.
8. Balasundaram, G. and Webster, J. T. (2006) Nanotechnology and biomaterials for orthopedic medical applications, *Nanomedicine*, **1**(2), 169–176.
9. Balík, K., Sochor, M., Pešáková, V., Křena, J., Glogar, P. and Grogor, J. (1999) Carbon-carbon and carbon-polymer composite materials as components implants for bone surgery, *Acta Montana*, **113**, 57–62.
10. Bantignie, J. L. (2007) Carbon nanotubes composites, *3rd CANAPE Workshop on Carbon Nanotubes for Biomedical Applications*, Rome.
11. Belavoine, F., Schultz, P., Richard, C., Mallou, V., Ebbesen, T. W. and Mioskowski, C. (1999) Helical crystallization of proteins on carbon nanotubes: A first step towards the development of new biosensors, *Angew. Chem. Int. Ed.*, **38**(13–14), 1912–1915.
12. Bhargava, A. (1999) Nanorobots: medicine of the future, http://www.ewh.ieee.org/r10/Bombay/news3/page4.html
13. Blazewicz, M. (2001) Carbon materials in the treatment of soft and hard tissue injuries, *European Cells and Materials*, **2**, 21–29.
14. Busanelli, L., Squarzoni, S., Brizio, L., Tigani, D. and Sudanese, A. (1996) Wear in carbon fiber-reinforced polyethylene (poly-two) knee prostheses, *Chir. Organi. Mov.*, **81**, 263–267.
15. Cha, S. I., Kim, K. T., Lee, K. H., Mo, C. B. and Hong, S. H. (2005) Strengthening and toughening of carbon nanotube reinforced alumina anocomposite fabricated by molecular level mixing processes, *Scripta Materialia*, **53**, 793–797.
16. Chan, C. K., Kumar, T. S. S., Liao, S., Murugan, R., Ngiam, M. and Ramakrishnan S. (2006) Biomimetic nanocomposites for bone graft applications, *Nanomedicine* **1**, 177–188.
17. Chen, R. J., Zhang, Y., Wang, D. and Dai, H. (2001) Noncovalent sidewall functionalization of single-walled carbon nanotubes for protein immobilization, *J. Am. Chem. Soc.*, **123**(16), 3838–3839.
18. Cherukuri, P., Gannon, C. J. and Leeuw, T. K. (2006) Mammalian pharmacokinetics of carbon nanotubes using instrinsic near-infrared fluorescence, *Proc. Natl. Acad. Sci. USA*, **103**(50), 18882–18886.
19. Christenson, E. M., Anseth, K. S. van den Beucken, J. J. J. P., Chan, C. K., Ercan, B., Jansen, J. A., Laurencin, C. T., Li, W-L., Murugan, R., Nair, L. S., Ramaksihna, S., Tuan, S. T., Webster, J. T. and Mikos, A. G. (2006) Nanobiomaterial applications in orthopaedics, *J. Orth. Res.*, **10**, 11–22.

20. Clark, M. E., Farrar, D. F., Engel, C., Felstead, P. J. C., Walter, D. M., Walter, G. S., Scotchford, C. A. and Grant, D. M. (2006) *In vitro* evaluation of osteoblast and fibroblast response to nano-phase fillers, *Trans. Soc. Biomat. Meeting 2006*, **148**.
21. Collins, P. G. and Avouris, P. (2000) Nanotubes for electronics, *Scientific American*, December, 62–69.
22. Correa-Duarte, M. A., Wagner, N., Rojas-Chapana, J., Morsczeck, C., Thie, M. and Giersig M. (2004) Fabrication and biocompatability of carbon nanotube-based 3d networks as scaffolds for cell seeding and growth, *Nano. Lett.*, **4**(11), 2233–2236.
23. Dandy, D. J., Flanagan, J. P. and Steenmeyer, V. (1982) Arthroscopy And the anagement of the ruptured anterior cruciate ligamen, *Clin. Orth. Rel. Res.*, **167**, 43–49.
24. Dannemair, W. C., Haynes, D. W. and Nelson, D. L. (1985) Granulomous reaction and cystic bony destruction associated with high wear rate in a total knee prosthesis, *Clin. Orth. Rel. Res.*, **198**, 224–230.
25. Davidson, R., Brabon, S., Lee, R. J. and Nelson, K. (1998) The development of CFRP based hip stems: realising their commercial potential, *Trans. Eu. Conf. Composite Materials*, 513–518.
26. Davis J. R. (2003) *Handbook of Materials for Medical Devices. ASM International*
27. Dell'Acqua-Bellavitis, L. M., Ballard, J. D., Bizios, R. and Siegel, R. W. (2005) Nanotube probes for electrophysiological applications, *Trans. 30th Soc. Biomater. Meeting*, p. 498.
28. Deng X., Jia, G., Wang, H., Sun, H., Wang, X., Yang, S., Wang T. and Liu, Y. (2007) Translocation and fate of multi-walled carbon nanotubes *in vivo*, *Carbon*, **45**(7), 1419–1424.
29. Dubin, R. A., Callegari, G. C., Kuppler, J., Kornev, K. G., Ruetsch S., Neimark, A. V. and Kohn, J. (2006) Cell growth on single wall carbon nanotube fibers, *Trans. 31st Soc. Biomater. Meeting*, p. 310.
30. Elias, K. E., Price, R. L. and Webster, T. J. (2002) Enhaced function of osteoblasts on nanometer diameter carbon fibres, *Biomaterials*, **23**, 3279–3287.
31. Emami, N. (2007) Manufacturing of bionano-composites: CNT reinforced UHMWPE composite, *Abstr. Nanotech Northern Europe 2007*.
32. Evans, G. R. D. (2001) Peripheral nerve injury: A review and approach to tissue engineered constructs, *Anat. Record*, **263**, 396–404.
33. Evans, S. L. and Gregson, P. J. (1998) Composite technology in load-bearing orthopaedic implants, *Biomaterials*, **19**, 1329–1342.
34. Fearon, P. K., Watts, C. P., Hsu, W. K., Billingham, N. C., Kroto, H. W. and Walton, D. R. M. (2002) Impact of carbon nanotube addition on the oxidative stability of polyolefins, *Abstr. NanoteC02 — Nanotechnology in Carbon and Related Materials*.
35. Fiorito, S., Serafino, A., Andreola, F. and Bernier, P. (2006a) Effects of fullerenes and single-wall carbon nanotubes on murine and human macrophages, *Carbon*, **44**(6), 1101–1106.
36. Fiorito, S., Serafino, A., Andreola, F., Togna, A. and Togna, G. (2006b) Toxicity and biocompatibility of carbon nanoparticles, *J. Nanosci. Nanotech.*, **6**(3), 591–599.

37. Firkowska, I., Olek, M., Pazos-Peréz, N., Rojas-Chapana, J. and Giersig, M. (2006) Highly ordered mwnt-based matrixes: Topography at the nanoscale conceived for tissue engineering, *Langmuir*, **22**(12), 5427–5434.
38. Firth, A., Aggeli, A., Burke, J.L., Yang, X. and Kirkham, J. (2006) Biomimetic self-assembling peptides as injectable scaffolds for hard tissue engineering, *Nanomedicine*, **1**, 189–199.
39. Florian, H., Gojny, J. N., Zbigniew, R. and Schulte, K. (2003) Surface modified multiwalled carbon nanotubes in CNT/Epoxy-composites, *Chemical Physics Letters*, **370**, 820–824.
40. Forster, I. W., Ralis, Z. A., McKibbin, B. and Jenkins, D. H. R. (1978) Biological reaction to carbon fiber implants, *Clin. Orthop. Rel. Res.*, **131**, 299–307.
41. Gabay, T., Jakobs, E., Ben-Jacob, E. and Hanein, Y. (2005) Engineered self-organization of neural networks using carbon nanotube clusters, *Physica A*, **250**, 611–621.
42. Gabay, T., Ben-David, M., Kalifa, I., Sorkin, R., Abrams, Z. R., Hanein, Y. and Ben-Jacob, E. (2007) Electro-chemical properties of carbon nanotube based multi-electrode arrays, *Nanotechnology*, **18**, 1–6.
43. Goodman, C. M, McCusker, C-D., Yilmaz, T. and Rotello, V. M. (2004) Toxicity of gold nanoparticles functionalised with cationic and anionic side chains, *Bioconjug. Chem.*, **15**, 897–900.
44. Haddon, R. C., Apparao, M. R. and Mattson, M. P. (2003) Molecular functionalisation of carbon nanotubes and use as substrates for neuronal growth, United States Patent 6'670'179. University of Kentucky Research Foundation.
45. Halcomb, F. J. and Bardos, D. (1981) Carbon fiber-reinforced polyethylene (CPE) for total knee replacement prostheses: A clinical experience, *Trans. Am. Soc. Artif. Intern. Organs*, **27**, 364–368.
46. Harrison, B. S. and Atala, A. (2007) Carbon nanotube applications for tissue engineering, *Biomaterials*, **28**, 344–353.
47. Harutyunyan, A. R., Pradhan, B. K., Sumanasekera, G. U., Korobko, E. Y. and Kuznetsov, A. A. (2002) Carbon nanotubes for medical applications, *European Cells and Materials*, **3**(2), 84–87.
48. Hillebrenner, H., Buyukserin, F., Stewart, J. D. and Martin, C. R. (2006) Template synthesized nanotubes for biomedical delivery applications, *Nanomedicine*, **1**, 39–50.
49. Hoet, P. H., Bruske-Hohlfeld, I. and Salata, O. V. (2004) Nanoparticles–known and unknown health risks, *J. Nanobiotechnology*, **2**, 12–14.
50. Holzinger, M., Steinmetz, J., Samaille, D., Glerup, M., Paillet, M., Bernier, P., Ley, L. and Graupner, R. (2004) [2+1] cycloaddition for cross-linking SWCNTs, *Carbon*, **42**, 941–943.
51. Hong, S. H., Cha, S., Kim, K. T. and Hong, S. H. (2004a) Method of producing metal nanocomposite powder reinforced with carbon nanotubes and the power prepared thereby, United States Patent 7'217'311. Korea Advanced Institute of Science and Technology.

52. Hong, S. H., Cha, S., Kim, K. T., Lee, K. H. and Mo, C. B. (2004b) Ceramic nanocomposite powders reinforced with carbon nanotubes and their fabrication process. United States Patent 20'040'217'520. Korea Advanced Institute of Science and Technology.
53. Howard, C. B., Taylor, K. J. J. and Gibbs, A. (1985) The response of human tissue to carbon reinforced epoxy resin, *J. Bone Joint Surg.*, **67–B**, 656–658.
54. Hu, H., Ni, Y. C., Montana, V., Haddon, R. C. and Parpura, V. (2004) Chemically functionalised carbon nanotubes as substrates for neuronal growth, *Nano. Lett.*, **4**(3), 507–511.
55. Iijima, S. (1991) Helical microtubules of graphite carbon, *Nature*, **354**, 56–58.
56. Jenkins, D. H. R. (1985) Ligament induction by filamentous carbon fibre, *Clin. Orthop. Rel. Res.*, **197**, 86–90.
57. Jockisch, K. A., Brown, S. A., Bauer, T. W. and Merritt, K. (1992) Biological response to chopped-carbon-fiber-reinforced PEEK, *J. Biomed. Mater. Res.*, **26**, 133–146.
58. Jones, E., Streicher, R. M., Field, R. and Rushton, N. (2001a) Evaluation of the role of carbon fibre composite in acetabular components, *Trans. 5th EFORT 2001*, O **669**, p. 119.
59. Jones, E., Wang, A. and Streicher, R. M. (2001b) Validating the limits for a PEEK composite as an acetabular wear surface, *Trans. 27th Soc. Biomat. Meeting*, p. 488.
60. Kam, N. W., O'Connel, M. O., Wisdom, J. A. and Dai, H. (2005) Carbon nanotubes as multifunctional biological transporters and near-infrared agents for selective cancer cell destruction, *Proc. Natl. Acad. Sci. USA*, **102**(33), 11600–11605.
61. Kawaguchi, M., Fukushima T., Hayakawa, T., Nakashima, N., Inoue, Y., Takeda, S., Okamura, K. and Taniguchi, K. (2006) Preparation of carbon-nanotube alginate nanocomposite gel for tissue engineering, *J. Dent. Mater.*, **25**, 719–725.
62. Kennel, E. B. and Glasgow, D. G. (2001) Carbon nanofiber composites for aerospace, *Proc. NASA Nanospace 2001 — Exploring Interdisciplinary Frontiers*, 220.
63. Khang, D., Sato, M. and Webster, T. J. (2005) Directed osteoblast functions on micro-aligned patterns of carbon nanofibers on a polymer matrix, *Rev. Adv. Mater. Sci.*, **10**, 205–208.
64. Khang, D., Durbin, S. and Webster, T. J. (2006) Enhanced fibronection adsorption on carbon nanotubes in polycarbonate urethane compostes directs osteoblast adhesion, *Trans. 31st Soc. Biomat. Meeting*, p. 322.
65. Khare, R. and Bose, S. (2005) Carbon Nanotube Based Composites — A Review, *J. Minerals & Materials Characterization & Engineering*, **4**(1), 31–46.
66. Kotwal, A. and Schmidt, C. E. (2001) Electrical stimulation alters protein adsorption and nerve cell interactions with electrically conducting biomaterials, *Biomaterials*, **22**, 1055–1064.
67. Kroto, H. V., Heath, J. R., O'Brien, S. C., Curl, R. F. and Smalley, R.E. (1985) C_{60}: Buckminster–fullerene, *Nature*, **318**, 162–163.
68. Lancaster, J. K. (1986) Lubrication of carbon fiber-reinforced polymers. In *Friction and Wear in Composites*, Friedrich K. (Ed.). Elsevier, Amsterdam, pp. 363–370.

69. Lau Kin-Tak, D. H. (2002) The revolutionary creation of new advanced materials - carbon nanotube composites, *Composites Part B: Engineering*, **33**, 263–277.
70. Leach, D. C. and Moore, D. R. (1985) Toughness of aromatic polymer composites reinforced with carbon fibres, *Composites Science and Technology*, **23**, 131–161.
71. Lee, D., Kim, K. and Lee, M. (2005) Effect of electrochemical interaction between nafion and carbon nanotube on nanotube bundle size of multiwalled carbon nanotube dispersed ionomeric nanocomposites as biomimetic artificial muscles, *Trans. 30th Soc. Biomater. Meeting*, p. 501.
72. Lee, J. L., Kim, J. Y., Khang, D. and Webster, T. J. (2006) Stem cell impregnated carbon nanofibers/nanotubes for treating neurological disorders: An *in vivo* Study, *Trans. 31st Soc. Biomat. Meeting*, p. 150.
73. Levenston, M. E., Beaupre, G. S., Schurman, D. J. and Carter, D. R. (1993) Computer simulations of stress-related bone remodelling around non-cemented acetabular components, *J. Arthroplasty*, **8**, 595–60.
74. Liao, K. and Reifsnider, K. L. (1993) Multiaxial fatigue and life prediction of composite hip prostheses, *ASTM STP*, **1191**, 429–449.
75. Liopo, A. V., Stewart, M. P., Hudson, J., Tour, J. M. and Pappas, T. C. (2005) Biocompatibility of native and functionalised single-walled carbon nanotubes for neuronal interfaces, *J. Nanosci. Nanotechnol.*, **6**(5), 1365–1374.
76. Liu, H. and Webster, T. J. (2007) Nanomedicine for implants: A review of studies and necessary experimental tools, *Biomaterials*, **28**(2), 354–369.
77. Liu, Z., Cai, W., He, L., Nakayama, N., Chen, K., Sun, X., Chen X. and Dai, H. (2007) *in vivo* biodistribution and highly efficient tumor targeting of carbon nanotubes in mice, *Nature Nanotechnology*, **2**, 47–52.
78. Lovat, V., Pantarotto, D., Lagostena, L., Cacciari, B., Grandolfo, M., Righi, M. and Prato M. (2005) Carbon nanotube substrates boost neuronal electrical signaling, *Nano. Lett.*, **5**(6), 1107–1110.
79. MacDonald R. A. and Stegemann, J. P. (2006) Electrically conductive biopolymers incorporating carbon nanotubes, *Trans. 31 Soc. Biomater. Meeting*, p. 326.
80. MacDonald R. A., Laurenzi, B. F., Viswanathan, G., Ajayan, P. M. and Stegemann, J. P. (2005) Collagen–carbon nanotube composite materials as scaffolds in tissue engineering, *J. Biomed. Mater. Res.* A, **74**, 489–496.
81. Magee, F. P., Allan, D. V. M., Weinstein, M., Longo, J. A., Koeneman, J. B. and Yapp, R. A. (1988) A canine composite femoral stem, *Clin. Orthop. Rel. Res.*, **235**, 237–252.
82. Matsudai, M. and Hunt, G. (2005) Nanotechnology and public health, *Nippon Koshu Eisei Zasshi*, **52**, 923–927.
83. Mattson, M. P., Haddon, R. C. and Rao, A. M. (2000) Molecular functionalisation of carbon nanotubes and use as substrate for neuronal growth, *J. Mol. Neurosci.*, **14**, 175–182.
84. McKenzie, J. L., Waid, M. C., Shi, T. and Webster, T. J. (2004) Decreased functions of astrocytes on carbon nanofiber materials, *Biomaterials*, **25**, 1309–1317.

85. Miller, J. H. (1984) Comparison of the structure of neotendons induced by implantation of carbon or polyester fibres, *J. Bone Joint Surg.*, **66**, 131–139.
86. Mo, C. B., Cha, S. I., Kim, K. T., Lee, K. H. and Hong, S. H. (2004) Fabrication of carbon nanotube reinforced alumina matrix nanocomposite by sol-gel process, *Mater. Sci. Eng.* A, **395**, 124–128.
87. Moore, R., Beredjiklian, P., Rhoad, R., Thiess, S., Cuckler, J., Duchyene, P. and Baker, D. G. (1997) A comparison of the inflammatory potential of particulates derived from two composite materials, *J. Biomed. Mater. Res.*, **34**(2), 137–147.
88. Morrison, C., Macnair, R., MacDonald, C., Wykman, A., Goldie, I. and Grant, M. H. (1995) *in vitro* biocompatibility testing of polymers for orthopaedic implants using cultured fibroblasts and osteoblasts, *Biomaterials*, **16**, 987–992.
89. Mukherjee, D. P. and Saha S. (1993) The application of new composite materials for total joint arthroplasty, *J. Long Term Eff. Med. Implants*, **3**, 131–141.
90. Mylvaganam, K. and Zhang, L. C. (2007) Fabrication and application of polymer composite comprising carbon nanotubes, *Recent Patents On Nanotechnology*, **1**, 59–65.
91. Namilae, S., Chandra, N. and Shet, C. (2004) Mechanical behaviour of functionalised nanotubes, *Chemical Physics Letters*, **387**, 247–252.
92. Ni, Y., Hu, H., Malarkey, E. B., Zhao, B., Montana, V., Haddon, R. C. and Parpura, V. J. (2005) Chemically functionalised water soluble single-walled carbon nanotubes modulate neurite outgrowth, *J. Nanosci. Nanotechnol.*, **5**(10), 1707–1712.
93. Qian, D., Andrews, R. and Rantell, R. T. (2000) Load transfer and deformation mechanisms in carbon nanotube-polystyrene composites, *Applied Physics Letters*, **76**(20), 2868–2870.
94. Pace, N., Spurio, S., Pavan, L., Rizzuto, G. and Streicher, R. M. (2002) Clinical trial of a new CF-PEEK acetabular insert in hip arthroplasty, *Hip International*, **12**(2), 212–214.
95. Peppas, N. A., Hilt, J. Z. and Thomas, J. B. (2007) *Nanotechnology in Therapeutics: Current Technology and Applications*. Horizon Press, Norfolk.
96. Peterson, C. D., Hillberry, B. M. and Heck, D.A. (1988) Component wear of total knee prostheses using Ti-6A1-4V, titanium nitride coated Ti-6A1-4V, and cobalt-chromium-molybdenum femoral components, *J. Biomed. Mater. Res.*, **22**(10), 887–903.
97. Pienkowski, D. A. and Andrews, R. J. (2001) Polymethylmethacrylate augmented with carbon nanotubes, United States Patent 6'872'403. University of Kentucky Research Foundation.
98. Polineni, V. K., Wang, A., Essner, A., Lin, R., Chopra, A., Stark, C. and Dumbleton, J. H. (1998) Characterization of carbon fibre reinforced PEEK composite for use as a bearing material in total hip replacements, *ASTM STP*, **1346**, 266–273.
99. Price, R. L., Waid, M. C., Haberstroh, K. M. and Webster, T. J. (2003) Selected bone cell adhesion on formation of formulations containing carbon nanofibres, *Biomaterials*, **24,** 1877–1887.

100. Price, R. L., Ellison, K., Haberstroh, K. M. and Webster, T. J. (2004) Nanometer surface roughness increases selected osteoblast adhesion on carbon nanofiber compacts, *J. Biomed. Mater. Res.* A, **70**, 129–138.
101. Pryor, G. A., Villar, R. N. and Coleman, N. (1992) Tissue reaction and loosening of carbon-reinforced polyethylene arthroplasty, *J. Bone Joint Surg.*, **74-B**, 156–157.
102. Qian, D., Andrews, R. and Rantell, R. T. (2000) Load transfer and deformation mechanisms in carbon nanotube-polystyrene composites, *Appl. Phys. Lett.*, **76**(20), 2868–2870.
103. Sahoo, S. K., Parveen, S. and Panda, J. J. (2007) The present and future of nanotechnology in human health care, *Nanomedicine: Nanotechnology, Biology, Medicine*, **3**, 20–31.
104. Salata, O. (2004) Applications of nanoparticles in biology and medicine, *J. Nanobiotechnology*, **2**(1), 3–9.
105. Scholes, S. C., Unsworth, A., Streicher, R. M. and Jones, E. (2002) Preliminary testing of alternative bearing material combinations for a total knee prosthesis, *Trans. Europ. Soc. Biomat. Meeting*, p. 124.
106. Seal, B. L., Otero, T. C. and Panitch, A. (2001) Polymeric biomaterials for tissue and organ regeneration, *Mater. Sci. Eng.*, **R 34**, 147–230.
107. Shaffer, M. S. P., Sandler, J., Werner, P., Altstädt, V., van Es, M. and Windle, A. (2002) Nanofibre-reinforced melt-spun fibre and injection-moulded peek composites, *Abstr. NanoteC02 - Nanotechnology in Carbon and Related Materials.*
108. Shard, A. G. and Tomlins, P. E. (2006) Biocompatibility and the efficacy of medical implants, *Regenerative Medicine*, **1**(6), 789–800.
109. Shi, X., Hudson, J., Spicer, P. P., Krishnamoorti, R., Tour, J. M. and Mikos, A. G. (2005) Rheological behavior and mechanical reinforcement of poly(propylene fumarate)-based single-walled carbon nanotube composites, *Trans. 30th Soc. Biomater. Meeting*, p. 66.
110. Shi, X., Hudson, J., Spicer, P. P., Krishnamoorti, R., Tour, J. M. and Mikos, A. G. (2005) Rheological behavior and mechanical characterization of injectable poly(propylene fumarate)/single-walled carbon nanotube composites for bone tissue engineering, *Nanotechnology*, **16**, S531–S538.
111. Singh, R., Pantarotto, D., Lacerda, L., Pastorin, G., Klumpp, C. and Prato, M. (2006) Tissue biodistribution and blood clearance rates of intravenously administered carbon nanotube radiotracers, *Proc. Natl. Acad. Sci. USA*, **103**(9), 3357–3362.
112. Sirdeshmukh, R., Teker, R. and Panchapakesan, B. (2004) Biological functionalization of carbon nanotubes, *Symp. Proc. Mater. Res. Soc.*, p. 823
113. Skinner, H. B. (1988) Composite technology for total hip arthroplasty, *Clin. Orthop. Rel. Res.*, **235**, 224–236.
114. Sorkin, R., Gabay, T., Blinder, P., Baranes, D., Ben-Jacob, E. and Hanein, Y. (2006) Compact self-wiring in cultured neural networks, *J. Neural. Eng.*, **3**(2), 95–101.
115. Streicher, R. and Wang, A. (2003) Mechanical and tribological properties of crosslinked UHMWPE, *Rivista di patologia dell' apparato locomotivo*, **II**(1/2), 39–46.

116. Streicher, R. M., Schmidt, M. and Fiorito S. (2006) Nanostructures and nanosurfaces for artificial orthopedic implants, *Nanomedicine*, **2**(6), 861–874.
117. Supronowicz, P. R., Ajayan, P. M., Ullmann, K. R., Arulanandam, B. P., Metzger, D. W. and Bizios, R.G. (2002) Novel current-conducting composite substrates for exposing osteoblasts to alternating current stimulation, *J. Biomed. Mater. Res.*, **59**(3), 499–506.
118. Tanaka, K., Tamura, J., Kawanabe, K., Nawa, M., Oka, M., Uchida, M., Kokubo, T. and Nakamura, T. (2002) Ce-TZP/Al2O3 nanocomposite as a bearing material in total joint replacement, *J. Biomed. Mater. Res. (Appl Biomater)*, **63**, 262–270.
119. Thostenson, E. T., Ren, Z. and Chou, T.-W. (2001) Advances in the science and technology of carbon nanotubes and their composites: A review, *Composites Science and Technology*, **61**(13), 1899–1912.
120. Thostenson, E. T., Li, C. and Chou, T.-W. (2005) Nanocomposites in context, *Composites Science and Technology*, **65**(3–4), 491–516.
121. Thuaire, A., Goujon, C., Gauvin, R. and Drew, R. A. L. (2004) Study on the fabrication of aluminum matrix nanocomposites reinforced with carbon nanotubes, *Microscopy and Microanalysis*, **10**, 574–575.
122. Vander Sloten, J., Labey, L., Van Audekercke, D., Helsen, J. and Van der Perre, G. (1993) The development of a physiological hip prosthesis: The influence of design and materials, *J. Biomed. Mater. Eng.*, **4**, 1–10.
123. Vincent, H. L. and Lee, Y. (2004) Nanotechnology: Challenging the limit of creativity in targeted drug delivery, *Advanced Drug Delivery Reviews*, **56**(11), 1527–1528.
124. Wang, S.-F., Shen, L., ZhangW.-D. and Tong, Y.-J. (2005) Preparation and mechanical properties of chitosan/carbon nanotubes composites, *Biomacromolecules*, **6**, 3067–3072.
125. Webster, T. J., Waid, M. C., McKenzie, J. L., Price, R. L. and Ejiofor, J. U. (2004) Nanobiotechnoloogy. Carbon nanofibres as improved neural and orthopaedic implants, *Nanotechnology*, **15**, 48–54.
126. Webster, T. J. and Ah, E. S. (2006) Nanostructured biomaterials for tissue engineering bone, *Adv. Biochem. Engin./Biotechnol.*, **103**, 275–308.
127. Webster, T. J. and Ergun, C., Doremus, R. H., Siegel, R. W. and Bizios, R. (2000) Specific proteins mediate enhanced osteoblast adhesion on nanophase ceramics, *J. Biomed. Mater. Res.*, **51**, 475–483.
128. Weisenberger, M. C., Andrews, R. and Rantell, T. (2007) Carbon nanotube polymer composite: Recent developments in mechanical properties. In *Physical Properties of Polymers Handbook*, Part VI. Mark JE (Ed.), Springer, NY, 585–598.
129. Wenz, L. M., Merritt, K., Brown, S. A., Moet, A. and Steffee, A. D. (1990) *in vitro* biocompatibility of polyetheretherketone and polysulfone composites, *J. Biomed. Mater. Res.*, **24**, 207–215.
130. Werner, P., Altstaed, V., Jaskulka, R., Jacobs, O., Sandler, J. K. W., Shaffer, M. S. P. and Windle, A. H. (2004) Tribological behaviour of carbon-nanofibre-reinforced PEEK, *Wear*, **257**, 1006–1014.

131. White, A. A., Best, S. M. and Kinloch, I. A. (2007) Hydroxyapatite-carbon nanotube composites for biomedical applications: A review, *International Journal of Applied Ceramic Technology*, **4**(1), 1–13.
132. Williams, D. (2003) Revisiting the definition of biocompatibility, *Medical Device Technology*, **14**(8), 23–24.
133. Wilson, M., Kannangara, K., Smith, G., Simmons, M. and Crane C. (2002) *Nanotechnology: Basic Science and Emerging Technologies*, CRC Press, Boca Raton.
134. Wolff, J. (1891) *Das Gesetz der Transformation der Knochen*. A. Hirschwald, Berlin.
135. Wright, T. M., Fukubayashi, T. and Burstein, A. H. (1981) The effect of carbon fiber reinforcement on contact area, contact pressure, and time-dependent deforemation in polyethylene tibial components, *J. Biomed. Mat. Res.*, **15**(5), 719–730.
136. Wright, T. M., Astion, D. J., Bansal, M., Rimnac, C. M., Green, T., Install, J. N. and Robinson, R. P. (1998) Failure of carbon fibre-reinforced Pe total knee components; A report of 2 cases, *J. Bone Joint Surg.*, **70-A**, 926–932.
137. Xie, J., Chen, L., Aatre, K. R., Srivatsan, M. and Varadan, V. K. (2006) Somatosensory neuron growth on functionalised carbon nanotube mats, *Smart Mater. Struct.*, **15**, N85–N88.
138. Zanello, L. P., Zhao, B., Hu, H. and Haddon, R. C. (2006) Bone cell proliferation on carbon nanotubes, *Nano. Lett.*, **6**(3), 562–567.
139. Zhan, G., Kuntz, J. D. and Mukherje, A. K. (2003a) Anisotropic thermal applications of composites of ceramics and carbon nanotubes, United States Patent 6'976'532. The Regents of the University of California.
140. Zhan, G. D., Kuntz, J. D., Wan, J. and Mukherjee, A. K. (2003b) Single-Wall carbon nanotubes as attractive toughening agents in alumina-based nanocomposites, *Nat. Mater.*, **2**, 38–42.
141. Zhang, D., Kandadai, M. A., Cech, J., Roth, S. and Curran, S. A. (2006) Poly(L-lactide) (PLLA)/multiwalled carbon nanotube (MWCNT) composite: Characterization and biocompatibility evaluation, *J. Phys. Chem. B: Condens. Matter. Mater. Surf. Interfaces Biophys.*, **110**(26), 12910–12915.
142. Zhang, G., Latour, R. A. Jr., Kennedy, J. M., Del Schutte, H. and Friedman, R. J. (1996) Long-term compressive property durability of carbon fibre-reinforced polyetheretherketone composite in physiological saline, *Biomaterials*, **17**, 781–789.
143. Zhao, B., Hu, H., Mandal, S. K. and Haddon, R. C. (2005) A bone mimic based on the self-assembly of hydroxyapatite on chemically functionalised single-walled carbon nanotubes, *Chem. Mater.*, **17**(12), 3235–3241.

5

Toxicity of Carbon-Nanotubes

Silvana Fiorito

There is evidence, to date, that nanomaterials have different effects on human and animal health from those of micron-sized materials. Typically nanomaterials possess specifical properties — chemical, optical, magnetic, biological — that may lead in the human body to a response that is different and not directly predicted from the chemical constituents and compounds. For example, even a traditionally inert compound, such as gold, may behave differently in the body when it is introduced as a nanomaterial. Therefore, the world's scientists, industries, and governments are beginning to take a critical look at nanotechnology and to develop a research agenda for addressing key issues related to the impact of nanotechnology on health and environment.

The group of people currently most affected by carbon-nanomaterials are the people in the workforce and the laboratories that generate and handle the materials. Thus, health scientists need to understand how these materials may impact health in the workplace or the laboratory, how they interact biologically in the body and what are the health effects resulting from toxicity, environmental exposure, and potential exposure routes: whether the material is inhaled, ingested or absorbed cutaneously.

Determining toxicity of Carbon-Nanotubes (CNTs) is complicated. The major issue complicating the assessment of toxicity of these particles is represented by their polyedric aspects and by the multiple factors affecting their physico-chemical behaviour. On one hand, the peculiar characteristics of these particles such as the surface coating, the presence of metal catalysts and/or graphite, the exposure to UV radiation, the dispersion properties, the tendence to deposit as aggregates, due to high Van der Waals forces, determine structure and function, and influence greatly their behaviour in a biological environment, making them more or less toxic.[26]

Another factor complicating the evaluation of toxicity is represented by the variability of the biological environment with which these nanostructures are challenged. The cell responses to foreign body materials are quite different depending on the cell type, the condition of the experiment and the cell-cell interactions. Moreover, species' differences complicate research because some species, such as rats, are more sensitive to particles than others, making it difficult to extrapolate the results to humans.

The toxicity of CNTs needs to be understood in the framework of the materials characterization. If scientists do not understand the material from a physical and chemical perspective, they cannot interpret exposure or toxicity measurements. For these reasons, the results of the studies performed until now, concerning the toxicity and biocompatibility of CNTs have been conflicting and uncertain.

The biological safety of carbon nanoparticles is an area of investigation still in its infancy, the literature to show studies on carbon nanoparticles and cell interactions starting only in the last two to three years.

In the last two years there has been an explosion of scientific contributes focusing on analysis of the biological behaviour, in terms of toxicity and biocompatibility, of CNTs. The assessment of cytotoxicity and inflammatory potential of CNTs has been performed by many *in vivo* studies on mammals tissues (mice, rabbits, guinea pigs) and by *in vitro* studies on several human cell types (macrophages, skin cells, embryo kidney cells, lung tumor cells, leukemia cells, mesothelioma cells, T and B lymphocytes) and mammalian cells [murine macrophages, adherent mammalian cells (A549, HeLa, Mod-K), mammalian cell suspension (Jurkat)]. Furthermore, other cells have been challenged with CNTs in order to determine if these particles can be internalized by different cellular species: fungal cells (Cryptococcus neoformans), yeast cells (*Saccharomices cerevisiae*), bacteria cells (*Escherichia coli*).

The results of these studies are contradictory and not uniform and show that many different factors both on the cell side and on the material side influence the behaviour of these particles towards the biological environment.

5.1. Behaviour and Fate of Carbon-Nanotubes in Mammals

At present, little is currently known about the behaviour and fate of CNTs in mammals, mostly because of the challenge of detecting and tracking these particles in complex biological environments. Parameters such as blood circulation and clearance half-life, organ biodistribution and accumulation have not been determined yet in an exhaustive way. Recently, Cherukuri P. *et al.*[2] measured the blood elimination kinetics of single-walled carbon nanotubes (SWNTs) in rabbits using their intrinsic near-infrared fluorescence, which is a property only of individualized SWNTs when excited with visible light at a wavelength between 900 and 1600 nm. They observed that the nanotubes

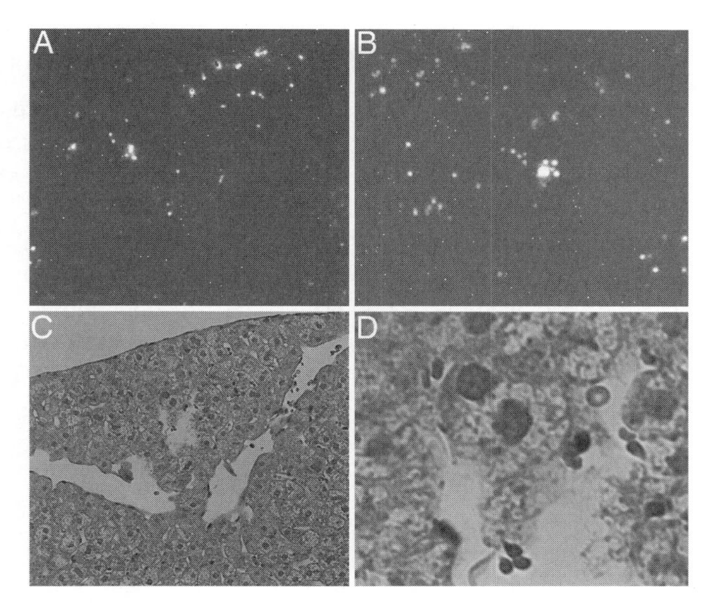

Fig. 1. Micrographs at two magnifications of liver tissue from rabbits killed 24 h after i.v. administration of suspended SWNTs. (A and B) Near-IR SWNT fluorescence images with field widths of 390 μm (A) and 83 μm (B). Scattered isolated bright pixels are artifacts from defective sensor elements in the near-IR camera; all larger features represent emission from SWNTs. In C and D, the SWNT fluorescence from A and B is shown overlaid as false-color green onto visible bright-field images from adjacent 3-μm-thick specimen slices that had been stained with hematoxylin and eosin. From, Cherukuri P *et al.*, *PNAS* (2006), **103**, 18882–18886.

concentration in the blood serum decreased exponentially with a half-life of 1.0 ± 0.1 h and that at 24 h after i.v. administration significant concentrations of nanotubes were found only in the liver (Fig. 1). The results reported reveal the behavior of nanotubes that remain disaggregated *in vivo*, as shown by their retention of near-IR fluorescence. During the 24-hour period between exposure and death, the experimental animals displayed normal behavior and no evidence of adverse effects from i.v. SWNT administration at the 20 μg/kg dosage used. In addition, pathological examination during necropsy, revealed no gross organ abnormalities, and histological evaluation of tissue sections found no pathological differences between the experimental and the control animals. Therefore, the authors deduce an absence of acute toxicity for the i.v. SWNT dosage used in the study. The localization of CNTs in the liver is shown in Fig. 1(A), that reveals numerous regions in the liver specimen with significant SWNT concentrations. The more magnified image of Fig. 1(B), shows one or two green clusters in addition to $\approx$30 diffraction-limited green spots. Each of these spots is emission from a single semiconducting SWNT. Similar analysis of tissue specimens from the kidneys, lungs, spleen, heart, brain, spinal cord, bone, muscle, pancreas, intestine, and skin, revealed far fewer or no nanotubes, and the control specimens showed no emissive features identifiable as SWNTs. The researchers conclude that, at 24 h after i.v. injection, the only significant SWNT concentration is in the liver.

Others[33] showed that intravenous administration of functionalised SWNTs (f-SWNTs) labelled with indium (^{111}In), followed by radioactivity tracing using gamma scintigraphy, indicated that f-SWNTs were not retained in any of the reticuloendothelial system organs (liver or spleen) and were rapidly cleared (half-life: 3 h) from systemic blood circulation through the renal excretion route. Furthermore, the same authors observed that urine excretion studies using both f-SWNTs and f-multiwalled CNT (f-MWNTs) followed by electron microscopy analysis of urine samples revealed that both types of CNT were excreted as intact nanotubes (Fig. 2).

5.2. Cellular Uptake of Carbon-Nanotubes

It has been demonstrated[13] that functionalised SWNTs and MWNTs exhibit a capacity to be taken up by a wide range of cells, such as, adherent mammalian cells, mammalian cell suspension, fungal cells (*Cryptococcus neoformans*), yeast cells (*Saccharomices cerevisiae*), bacteria cells (*Escherichia coli*) and can intracellularly traffic through different cellular barriers. F-CNTs were localised in the perinuclear region after 2 h of incubation with the cells. It could be also observed that the nature of the functional group on the CNT surface did not determine whether f-CNTs were internalized or not (Fig. 3).

5.3. Lung Toxicity of Carbon-Nanotubes

Humans have been exposed to nanoparticulates, especially natural environmental nanoparticulates, such as mineral species, combustion products and other anthropogenic nanoparticulates for millennia, if not longer. It has recently been showed that aggregated MWNTs and other fullerenic polyhedra are not only ubiquitous in the contemporary outdoor environment, but homes with gas cooking ranges (which are sources of these aggregates) may contain two orders of magnitude more of these aggregates than the ambient outdoor air.[19,21] Furthermore aggregates of MWNTs have been observed

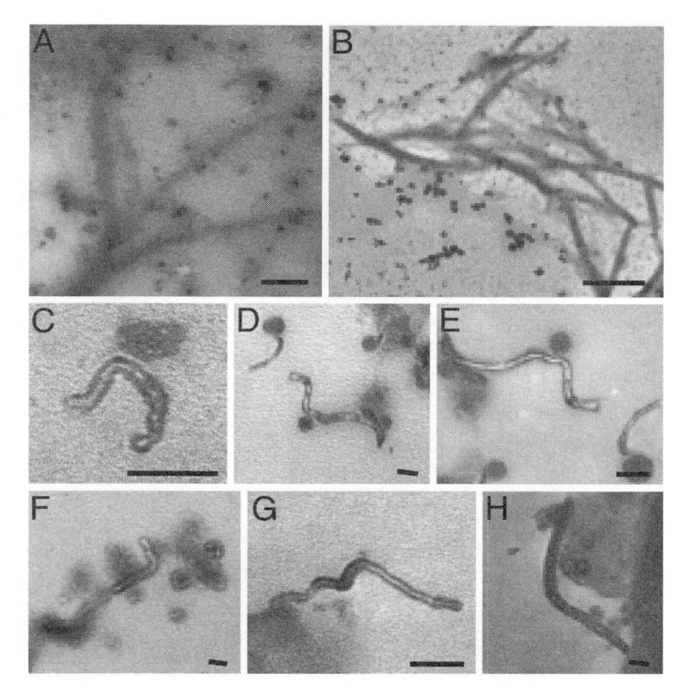

Fig. 2. TEM images of excreted urine samples containing single- and multiwalled DTPA–CNT. The urine samples were centrifuged, and both the supernatant and the precipitate were analyzed. (A and B) DTPA–SWNT from the supernatant. (Scale bars, 500 nm.) (C–E) DTPA–MWNT into the supernatant. (F–H) DTPA–MWNT in the precipitate. (Scale bars for C–H, 100 nm.) From, Singh R. *et al., PNAS* (2006), **103**, 3357–3362.

in a 10 000 year old Greenland ice core sample, indicative of the fact that these carbon nanoforms have been a component of the natural athmospheric combustion product in antiquity as well.[7] The same authors[20] performed viability assays on a murine lung macrophage cell line to assess the comparative cytotoxicity of commercial SWNTs and MWNTs. They observed that the commercial MWNTs aggregates, are identical in nanostructure and composition to those aggregates produced ubiquitously in the environment, and that, especially in indoor environments, where natural gas or propane cooking stoves exist, these aggregates create cytotoxic responses identical to those for chrysotile asbestos and black carbon aggregates. Clinical data for asthmatic patients in US show that 83% of women with mild to severe asthma are currently exposed to kitchen gas cooking and the 67% are continuously exposed. This leads the researchers to suggest that anthropogenic occurrence of CNT aggregates can contribute to allergies and/or asthma in humans for long term

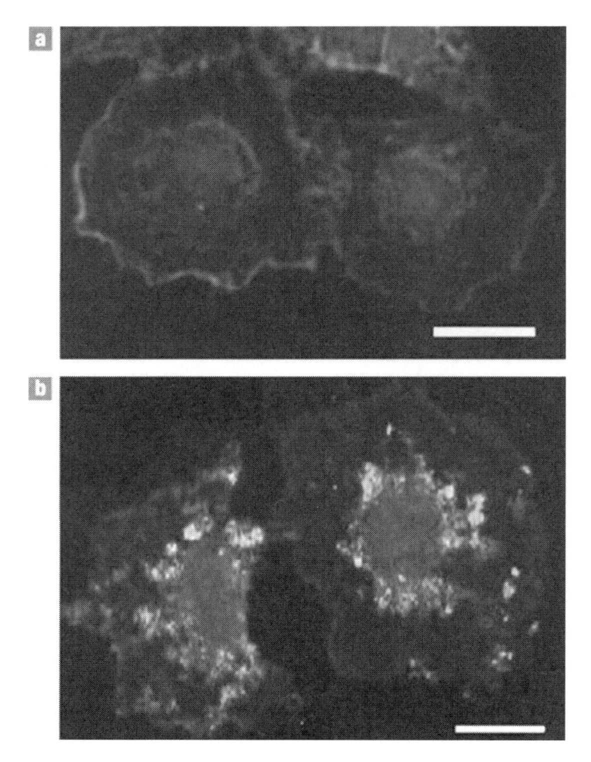

Fig. 3. Confocal image of mammalian cells showing the cellular uptake of CNTs From, K. Kostarelos *et al., Nature Nanotechnol.* (2007), **2**, 108.

exposure. These results raise concern also for environmental health effects of these carbon nanoparticulates, not previously considered either epidemiologically or etiologically, especially for the increasing incidence of respiratory diseases, particularly asthma, occurring in industrially more developed countries.

Most of the CNT cytotoxicity studies have focused on the pulmonary toxicity after inhalation, intratracheal instillation and pharyngeal aspiration, as well as their effects on skin toxicity after exposure and subcutaneous administration of CNTs. Huczko *et al.*[11] studied the effects of CNTs containing soot in intratracheally instilled guinea pigs, by evaluating pulmonary function parameters and bronchoalveolar lavage examination. They showed that after four weeks, a single intratracheal instillation of 25 mg CNTs containing soot, did not change pulmonary function and did not induce any measurable inflammation in bronchoalver space of the exposed guinea pigs. Warheit *et al.*[35]

evaluated the acute lung toxicity of intratracheally instilled single-wall carbon nanotubes. The lungs of rats were instilled with 1 or 5 mg/kg of SWNTs and were assessed using bronchoalveolar lavage, fluid biomarkers and cell proliferation methods, and by histopathological evaluation at 24 h, one week, one month and three months postinstillation. Exposure to high doses (5 mg/kg) of SWNTs produced mortality in ~15% of the instilled rats within 24 h postinstillation. This mortality resulted from mechanical blockage of the upper airways by the instillate and was not due to inherent pulmonary toxicity of the instilled SWNTs particulate. In the other animals, the intratracheal instillation exposure to SWNTs produced only a short term inflammatory and cell injury effects, as evidenced by a transient increase in the percentage of bronchoalveolar-recovered neutrophils, enzymes, and cell proliferation rates, at 24 h post exposure; but these values were not sustained through the other postexposure time periods. The lung histophatology study showed a non-dose-dependent series of multifocal granulomas, non uniform in distribution and not progressive, beyond 1-month post-exposure. On the contrary, much more severe lung lesions were observed in a study of Lam and colleagues[15] in which 23 mice were intratracheally instilled with 0, 0.1, 0.5 mg of carbon nanotubes, a carbon black negative control and a quartz positive control, and euthanized after 7 and 90 days for histopathological studies of the lungs. All nanotubes preparations induced dose-dependent epithelioid granulomas in the lungs and interstitial inflammation in the animals of the 7-day group. These lesions persisted and were more pronounced in the 90-day group. The lungs of some animals also revealed peribronchial inflammation and necrosis that were extended to the alveolar septa. The lungs of mice treated with carbon black were normal, whereas those treated with high-dose quartz revealed mild to moderate inflammation. The authors' conclusions are that, if CNTs reach the lungs, they are much more toxic than carbon black and can be more toxic than quartz. Another study showed that MWNTs and ground MWNTs, intratracheally administered to rats, were observed in the rat lungs after 60 days from the instillation, and that both induced inflammatory and fibrotic reactions characterized by the formation of collagen-rich granulomas protruding in the bronchial lumen, of alveolitis in the surrounding tissues and by the production of pro-inflammatory cytokines (TNF-alfa) *in situ*.[18] Also Shvedova *et al.*[32] demonstrated that pharyngeal aspiration of SWNTs elicited unusual pulmonary effects in mice that combined a strong acute inflammation with early onset of progressive fibrosis and granulomas (Fig. 4). Moreover, a dose-dependent increase in the markers of inflammation (LDH and gamma-GT) was found in bronchoalveolar lavage, as well as early elevation

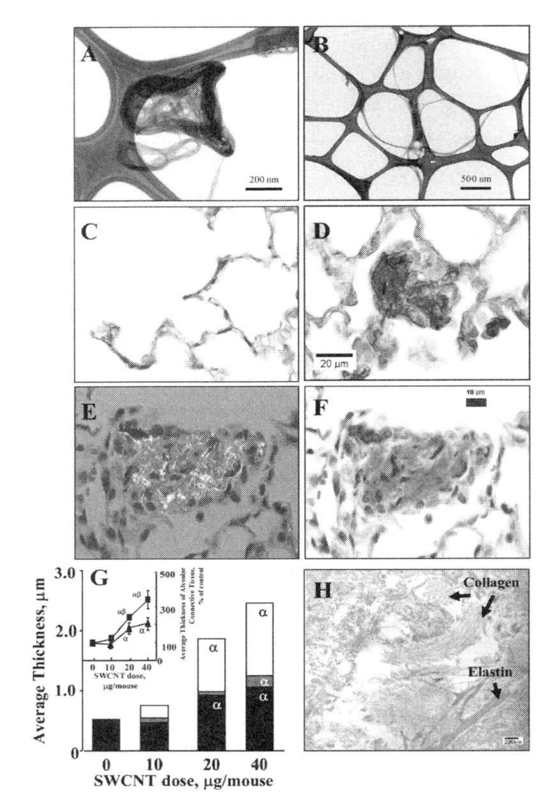

Fig. 4. (C) Lung sections seven days after aspiration of PBS (control mice). (D) Focal granulomatous inflammation seven days after aspiration of 40 μg SWCNT/mouse (E) Carbon nanotube material in a pulmonary granulomatous lesion 2 months after aspiration. (F) SWCNT are visualized as gray areas in the corresponding transmitted light section. From, Shvedova *et al., Am. J. Phisiol. Lung Cell Mol. Physiol.* (2005), **289**, L698.

of proinflammatory cytokines (TNF-alfa, IL-1beta) and pro-fibrogenic factors (TGF-beta).

Mangum *et al.*[15] investigated the fibrogenic potential of SWNTs following a single oropharyngeal aspiration in rats. No overt inflammatory response was observed at 1 or 21 days post-exposure as determined by histopathology and evaluation of cells in bronchoalveolar lavage fluid. However, SWNTs induced the formation of small focal interstitial fibrotic lesions within the alveolar region of the lung at 21 days. Interestingly, SWNTs were observed to form intercellular structures that bridged alveolar macrophages *in situ*.

5.4. Cell Toxicity of Carbon-Nanotubes

Human epidermal cells A study performed in human volunteers and in animals in order to assess the skin sensitivity to CNTs, showed that soot containing a high content of CNTs did not induce any skin irritation in fourty people subjected to a dermatological testing, or any eyes irritation in rabbits instilled with the soot.[10] On the contrary, Shvedova A.A. and colleagues[31] investigated adverse effects of SWCNTs, containing a certain amount of iron (30%), towards skin cells, using a cell culture of immortalized human epidermal keratinocytes. After 18 h of exposure they observed formation of free radicals, accumulation of peroxidative products, antioxidant depletion (decrease of intracellular levels of glutathione), oxydation of protein SH groups, and depletion of total antioxidant reserve and vitamin E. Additionally they observed decrease of cell viability as well as ultrastructural and morphological changes in cultured skin cells. Interestingly, an iron chelator dramatically reduced the cytotoxicity of SWNTs. The researchers concluded that SWNTs exposure can result in accelerated oxidative stress and may produce dermal toxicity in exposed workers. The induction of increased oxidative stress and inhibition of cell proliferation in human keratinocyte cells by SWNTs, was also observed by Manna *et al.*[16] In addition, these authors found that SWNTs activate NF-kappaB in a dose dependent manner, and that the mechanism of activation of NF-kappaB was due to the activation of stress-related kinases. On the contrary, others observed that CNTs could act as free radicals scavengers instead of inducing oxidative stress. It has been reported that purified MWNTs in aqueous suspension do not generate oxygen or carbon-centered free radicals in the presence of H_2O_2, but, when in contact with an external source of hydroxyl or superoxide radicals MWNTs exhibit a remarkable radical scavenging capacity towards reactive oxygen species, both HO and O_2, no matter how such species are generated.[9] Others[17] explored the effects of Multi-Walled-Nanotubes (MWNTs) on human epidermal keratinocytes, examining the cell uptake of these particles by transmission electron microscopy. They observed that, chemically, unmodified MWNTs were present within cytoplasmic vacuoles, and that, these induced the release of the proinflammatory cytokine interleukine 8 from the keratinocytes in a time dependent manner. In human epidermal keratinocytes exposed to MWNTs an altered expression of 36 proteins associated with metabolism, cell signalling and stress, was observed after 24 h of exposure, and of 106 proteins after 48 h, as well as, a consistent effect on the expression of cytoskeletal elements and vesicular trafficking components.[37] Another study showed in human epidermal keratinocytes, challenged *in vitro* with functionalised SWNTs, a significant decrease in cell viability and a dose-dependent increase in the production of pro-inflammatory cytokines (Il-6 and Il-8).[38]

Human lung cells The cellular toxicity of CNTs (MWNTs, Carbon-nanofibers, Carbon black) on human lung- tumor cell lines, was also studied as a function of their aspect ratio and surface chemistry.[14] MWNTs appeared to be less toxic than Carbon nanofibers and Carbon black particles and it was observed that the toxicity increased with the chemical surface treatment, when carbonyl, carboxyl and/or hydroxyl groups were present on their surface. So the authors concluded that the hazardous effect of CNTs is size-dependent and cytotoxicity is enhanced when the surface of the particles is functionalised after an acid treatment. A dose and time-dependent increase of intracellular reactive oxygen species and a decrease of the mitochondrial membrane potential was detected in human lung cells (A549) as well as in rat macrophages after CNTs (commercial, not purified SWNTs, and MWNTs) treatment, while the treatment with purified CNTs had no effect. This lead the authors to the conclusion that metal traces associated with the commercial nanotubes are responsible for the biological effects.[23] On the contrary, an *in vitro* cytotoxicity assessment of SWNTs on a human lung cell line (A549 cells) showed that SWNTs have a very low acute toxicity, and transmission electron microscopy (TEM) studies confirmed that there was no intracellular localization of SWCNT in A549 cells following 24 h exposure; but these particles were found to interfere with a number of the dyes used in the tests, so that no conclusive results could be done.[4]

Other cells Bottini[1] compared the toxicity of pristine-hydrophobic and oxidized MWNTs on human T cells and found that oxidized CNTs are more toxic and induce massive loss of cell viability through programmed cell death at doses of 400 μg/ml within five days. Pristine CNTs were less toxic than functionalised CNTs, but both pristine and oxidized CNTs were toxic in a dose and time-dependent manner. The toxicity of iron rich SWNTs (non-purified) and iron-stripped SWNTs (purified) has been studied on a human macrophage cell line in order to evaluate their capacity to induce oxidative stress. It was observed that non purified, iron-rich CNTs, were more effective in generating hydroxyl radicals by the cells and induced a major increase of the biomarkers of oxidative stress, enhancement of lipid peroxidation and anti-oxidant depletion.[12] These results, showing that iron-rich CNTs are more toxic than purified CNTs are in agreement with those of Shvedova and coworkers[31] who observed that an iron chelator induced a marked decrease in toxic effects of CNTs. The effects induced by CNTs on human macrophage cells have been also recently explored by Fiorito S. *et al.*[9] They aimed to assess the ability of pure Single-Walled-Carbon-Nanotubes (SWNTs) and C_{60}-fullerenes to elicit an inflammatory response by murine and human macrophage cells, and to be cytotoxic against these cells *in vitro*. Moreover, in order to determine if the behaviour of these C-nanoparticles towards the biological environment could be due, at least in part, to graphite, one of the main component of C-nanomaterial preparations, they compared the effects induced by carbon nanoparticles with those induced on the same cells by graphite particles. The following C-nanoparticles preparations were studied:

(a) highly purified SWNTs (containing only traces of graphite, nickel and cobalt, less than 1% weight) produced by the Chemical Vapour Deposition technique. (CVD) (Fig. 5);
(b) SWNTs produced by the electric arc technique;
(c) SWNTs, produced by the electric arc technique, cut and opened;
(d) C_{60}-fullerenes synthesized by the electric arc technique, and as reference:
    (1) synthetic highly purified graphite particles (Sigma) (diameter $= 1\mu$). They investigated:
        (i) the uptake of the above mentioned C-nanostructures by human monocytes-derived-macrophages (MDMs);

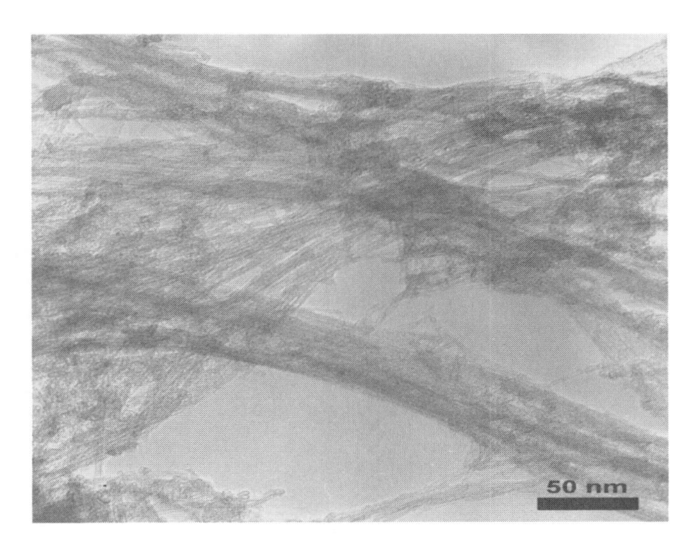

Fig. 5. TEM image of purified SWNTs. From, Fiorito S. *et al.*, *JNN* (2006), **6**(3), 591–599.

(ii) the cytotoxicity of these C-nanoparticles versus human MDMs;
(iii) the cell morphological changes induced by the contact with carbon nanoparticles.

Purified SWNTs, as well as C_{60}-fullerenes, were observed to induce apoptosis, cell death and metabolic alterations in a smaller percentage of cells as compared to both graphite particles and not pure and opened SWNTs (Figs. 6, 7). Purified SWNTs as well as C_{60}-fullerenes induced ultrastructural alterations, expression of activation-inflammation injury, in macrophage cells at a lesser extent than not pure particles (Fig. 8). The authors concluded that C-nanotubes, when highly purified and not containing graphite and catalysts, do not stimulate a reactive-inflammatory response by human cells *in vitro* and are less cytotoxic towards these cells than not pure SWNTs and graphite particles.

Furthermore, Salvador-Morales *et al.* studied the interaction between MWNTs and double-walled-nanotubes and a part of the human immune system, the complement cascade. The authors conducted haemolytic assays to investigate the activation of the human serum complement system via

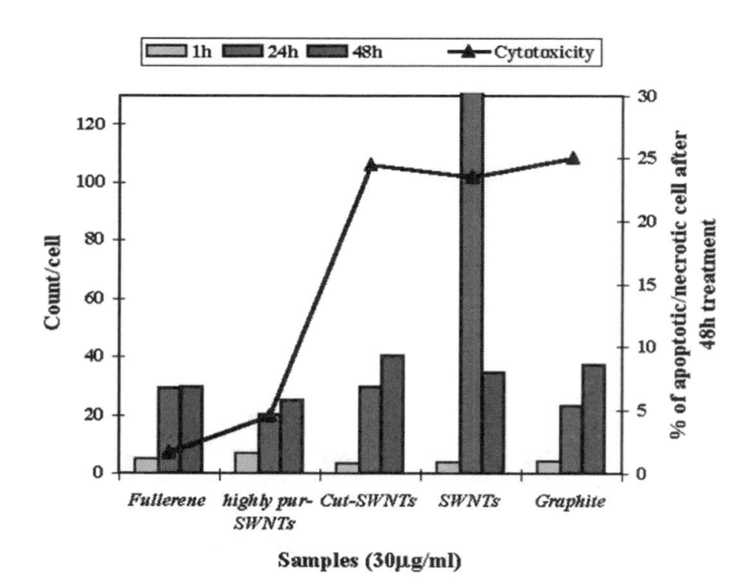

Fig. 6. Percentage of phagocytized C-nanoparticles (in columns) and of necrotic/apoptotic cells (black line) stained with propidium iodide and observed by CLSM. From, S. Fiorito *et al., JNN* (2006), **6**(3), 591–599.

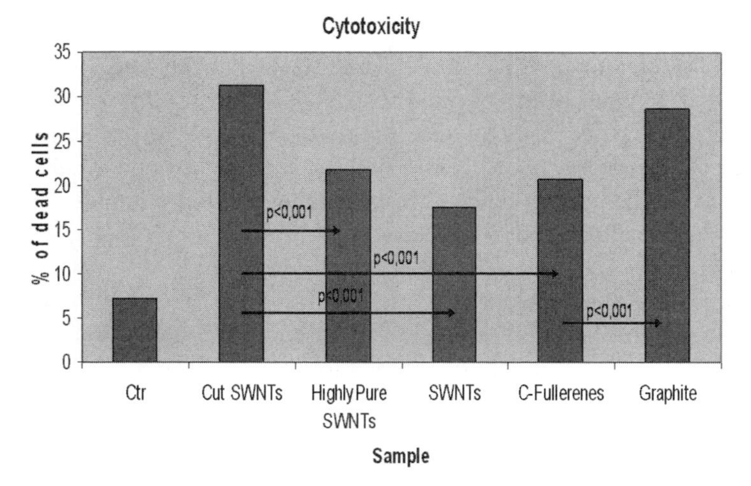

Fig. 7. Live/dead cell cytotoxicity test performed on human macrophage cells challenged with C-nanoparticles. From, Fiorito S. *et al.*, *JNN* (2006), **6**(3), 591–599.

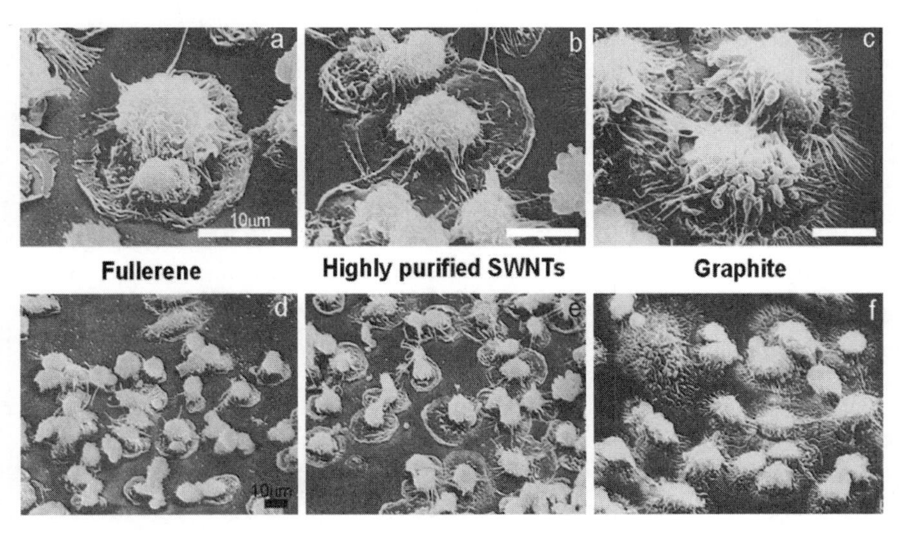

Fig. 8. SEM image of human macrophages challenged with carbon-nanoparticles. From, Fiorito S. *et al.*, *JNN* (2006), **6**(3), 591–599.

the classical and alternative pathways. They observed that carbon nanotubes activate human complement via both classical and alternative pathways and concluded that carbon nanotubes may promote damaging effects of excessive complement activation, such as inflammation and granuloma formation.

Recently, Cui[3] investigated the influence of single-walled carbon nanotubes (SWCNTs) on human HEK293 cells with the aim of exploring SWCNTs biocompatibility. Their results showed that SWCNTs can inhibit HEK293 cell proliferation, and decrease cell adhesive ability in a dose- and time-dependent manner. They demonstrated that SWNTs can inhibit human cells growth by inducing cell apoptosis and decreasing cellular adhesion ability. In a very recent study, researchers at Rice University's Center for Biological and Environmental Nanotechnology (CBEN) in Houston, Texas, have found that water soluble carbon nanotubes are significantly less toxic than those in their native state, which are insoluble.[30] The researchers exposed human dermal fibroblast cell cultures to varying doses of four types of water soluble SWNTs. These four included, pure undecorated SWNTs suspended in a soapy solution, and three forms of nanotubes that were rendered soluble via the attachment of the chemical subgroups hydrogen sulfite, sodium sulfite and carboxylic acid. It has been observed that the modified nanotubes were non-cytotoxic and that the cytotoxic response of cells in culture was dependent on the degree of functionalization of the single-walled carbon nanotubes (SWNTs). The authors found that as the degree of sidewall functionalization increases, the SWNT sample becomes less cytotoxic, and sidewall functionalized SWNT samples are substantially less cytotoxic than surfactant stabilized SWNTs.

The effect of length on CNTs toxicity has been investigated by Sato *et al.* who assessed the activation of the human monocytic cell line THP-1 *in vitro* and the response in subcutaneous tissue *in vivo* to MWNTs of different lengths. They used 220 nm and 825 nm long CNTs samples. Both 220-CNTs and 825-CNTs induced monocytic cells to produce a proinflammatory cytokine (TNF-alfa) in a dose-dependent manner, although their stimulating activity was much lower compared to that of a microbial antigen. In the *in vivo* test, the subcutaneous tissue inflammatory response around 220-CNTs was lower than that observed around 825-CNTs. Four weeks later the inflammatory response around 220-CNTs almost disappeared, while the degree of inflammation around 825-CNTs was the same, compared with the response observed after one week, and appeared like a foreign body granuloma. The authors concluded that the degree of inflammatory response in subcutaneous tissue in rats around 220-CNTs was slight in comparison with that around 825-CNTs since macrophages could envelop 220-CNTs more readily than 825-CNTs. However no severe inflammation response such as necrosis, degeneration or neutrophil infiltration *in vivo* was observed around both CNTs examined. The molecular characterization of the cytotoxic mechanism of MWNTs on human skin fibroblasts has been explored by Ding L *et al.* by phenotypic measurements, through the whole genome expression array analysis, of the cells exposed to MWNTs. They demonstrated that, exposing cells to MWNTs at cytotoxic doses (0.6 and 0.06 μg/ml) induces cell cycle arrest and increases apoptosis/necrosis. MWNTs exposure activates genes involved in cellular transport, metabolism, cell cycle regulation and stress response, that are indicative of a strong immune and inflammatory response within skin fibroblasts. Another study exploring the effects of SWNTs on stress genes in human cells showed induction of oxidative stress in these cells and increase in stress responsive genes.[28] The impact of two types of functionalised CNTs on cells of the immune system has been studied by Dumortier[6] who found that both types of f-CNTs are uptaken by B and T lymphocytes as well as by macrophages *in vitro*, without affecting cell viability, but, the one more water soluble did not affect cell function, while that one with reduced solubility induced secretion of pro-inflammatory cytokines by macrophages. Thus suggesting that the kind of functionalisation can influence the cell inflammatory response.

The cytotoxicity of CNTs at various degrees of agglomeration has been also investigated, in order to find out how far the degree of dispersion and agglomeration affects CNTs cytotoxicity. The cytotoxic effects of well-dispersed CNTs were compared with that of conventionally purified rope-like agglomerated CNTs using asbestos as reference.[36] Four different SWNTs fractions (CNT-row material, CNT-agglomerates, CNT-bundles and CNT pellet) were tested in an *in vitro* cytotoxicity assay using a human mesothelioma cell line (MSTO-211H). All CNT fractions, except the well dispersed CNT bundles, were aggregated, after the incubation time, to micron-sized structures. CNT raw material was able to significantly decrease cell activity and proliferation in a dose-dependent way. The CNT-agglomerates were the most toxic fraction (Fig. 9). After three days of incubation with cell cultures the treatment with CNT-agglomerates evoked round-shaped cell morphology of the cells, very similar to the cells treated with asbestos. In contrast, CNT-bundles did not induce any visible changes of the treated cells. A similar adverse effect on cell morphology and viability was showed by the CNT-pellet fraction enriched by carbonaceous material. The CNT-pellet and the CNT-agglomerate fractions were more toxic than well-dispersed CNT, containing low carbonaceous material but the same content of metal catalyst. Thus, the results of this study suggest that critical features that seem to determine CNT toxicity are the presence of carbonaceous material and the degree of CNT dispersion. The content of catalyst metals into the CNT samples does not seem to be one of the major issues influencing the cytotoxicity of these nanostructures.

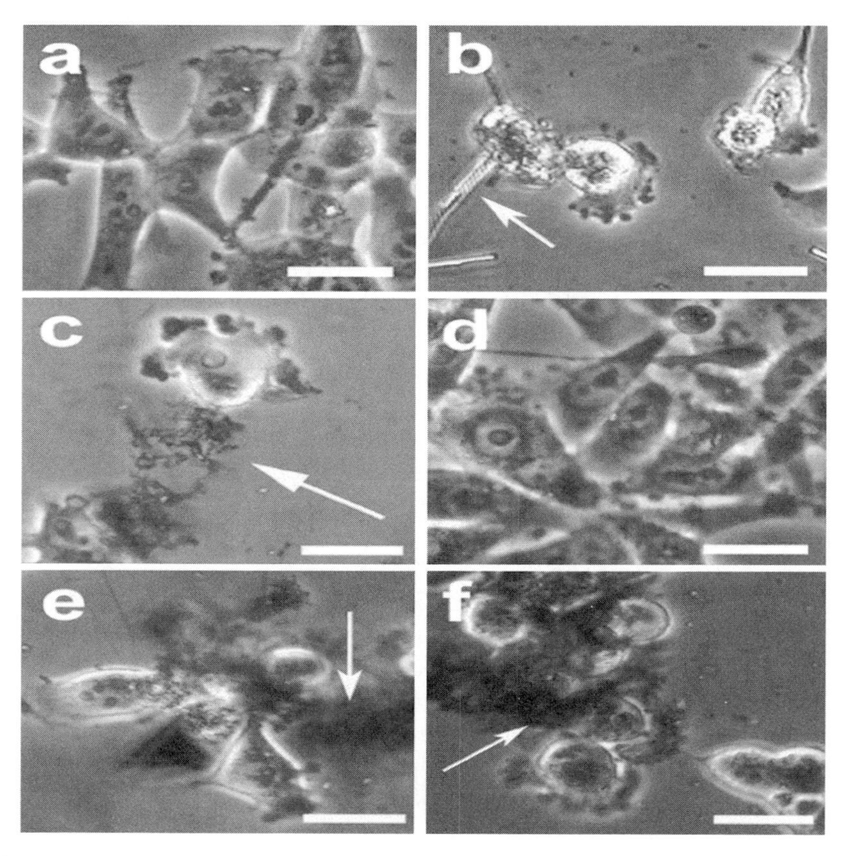

Fig. 9. Morphology changes of MSTO-211H cells after three days of exposure to 15 μg/ml of different fractions of CNTs and asbestos. (a) Untreated MSTO-211H cell culture and (b) cell culture exposed to asbestos. Arrows indicate needles of asbestos. (c) Cells treated with CNT-agglomerates were round-shaped and lost their adherence on the cell culture plate. Arrow points to CNT agglomeration. (d) Cells exposed to CNT-bundles showed no visible morphological changes compared to the control cells. (e) Effect of CNT-pellet fraction. Non-tubes material agglomerated during the incubation period to micro-sized structures. (f) Cells incubated with CNT raw material. Arrow indicates agglomerated CNT material: scale bar 20 μm. From, Wick P. *et al., Toxicology Letters* (2007), **168**, 121–131.

The effect of CNT on human platelet function *in vitro*, and rat vascular thrombosis *in vivo*, has been investigated by Radomski A. *et al.*[24] Incubation of platelets with MWNTs and SWNTs resulted in a concentration-dependent increase in platelet aggregation. Furthermore they observed *in vivo* that infusions of these CNTs significantly accelerated the time and rate of development of carotid artery thrombosis in rats. The authors observed that CNTs caused only partial aggregation with little or no platelet degranulation, and hypothesized that this effect could be due to the fact that, either carbon nanotubes can mimic molecular bridges involved in platelet-platelet interactions, thus stimulating aggregation, or that surface charges could play a role in direct interactions between these particles and platelet membrane receptors involved in aggregation mechanisms.

A toxicological assessment of five carbon nanomaterials on human fibroblast cells in vitro has been performed by Tian F. *et al.*[34] They correlated the physico-chemical characteristics of these nanomaterials (SWNTs, MWNTs, active carbon, carbon black, carbon graphite) to their toxic effects. They found that surface area is the variable that best predicts the potential toxicity of these purified nanomaterials, that purified SWNTs are more toxic than their unpurified counterpart, and that, for comparable small surface areas, dispersed carbon nanomaterials, due to a change in surface chemistry, induced cell morphological changes, cell detachment and apoptosis/necrosis. These results are in contrast with those of others[9,31,36] who found that CNTs, when purified and well dispersed are less toxic than those not purified and aggregated.

The cardiovascular effects of SWNTS were evaluated by the study of the oxidative stress induced aortic mitochondrial alterations. It was observed that a single intrapharyngeal instillation of SWNTs induced activation of markers of oxidative insults in lung, aorta and heart tissue in mice. Furthermore, mice exposed to SWNTs (10 and 40 μg/mouse) developed aortic mitochondrial DNA damage at 7, 28, and 60 days after exposure.[38] The impact of SWNTs on rat aortic smooth muscle cells was also examined by Raja[25] who observed a decrease in cell number and inhibition in cell growth over a 3-, 5-day time course. A dose dependent cytotoxicity against a fibroblast cell line (3T3) was also observed for three different SWNTs preparations (purified, unpurified and functionalized with glucosamine).[22]

To date, the studies exploring the cytotoxicity of CNTs, published mainly over the last three years, are not univocal and often contradictory. Based on their results we can just argue that the cytotoxic response to these nanoparticles is dependent on several different factors, among which the most relevant are: the degree of functionalisation, the degree of solubilization, the length of CNT, the size of the particles, the cell line, the tissue type, the material type, the degree and kind of agglomeration, the catalyst metal, the route of administration.

So many factors seem to influence the biocompatibility of CNTs that it is not possible, at present, to draw any consistent conclusion on their toxicity.

Acknowledgement

I am grateful to Mariangela Rasi for the assistance provided.

Bibliography

1. Bottini, M., Bruckner, S., Nika, K., Bottini, N., Bellucci, S., Magrini, A., Bergamaschi, A. and Mustelin, T. (2006) Multi-walled carbon nanotubes induce T lymphocyte apoptosis, *Toxicol. Lett.*, **160**, 121–126.
2. Cherukuri P., Gannon, C. J., Leeuw, T. K., Schmidt, H. K., Smalley, R. E., Curley, S. A. and Weisman, B. R. (2006) Mammalian pharmacokinetics of carbon nanotubes using intrinsic near-infrared fluorescence, *Proc. Nat. Acad. Sci. USA*, **103**, 18882–18886.
3. Cui, D., Tian, F., Ozkan, C. S., Wang, M., Gao, H. (2005) Effect of single wall carbon nanotubes on human HEK293 cells, *Toxicol. Lett.*, **155**, 73–85.
4. Davoren, M., Herzog, E., Casey, A., Cottineau, B., Chambers, G., Byrne, H. J. and Lyng, F. M. (2007) *In vitro* toxicity evaluation of single walled carbon nanotubes on human A549 lung cells, *Toxicol. In Vitro*, **21**, 438–448.
5. Ding, L., Stilwell, J., Zhang, T., Elboudwarej, O., Jiang, H., Selegue, J. P., Cooke, P. A., Gray, J. W. and Chen, F. F. (2005) Molecular characterization of the cytotoxic mechanism of multiwall carbon nanotubes and nano-onions on human skin fibroblast, *Nano Lett.*, **5**, 2448–2464
6. Dumortier, H., Lacotte, S., Pastorin, G., Marega, R., Wu, W., Bonifazi, D., Briand, J. P., Prato, M., Muller, S. and Bianco, A. (2006) Functionalized carbon nanotubes are non-cytotoxic and preserve the functionality of primary immune cells, *Nano Lett.*, **6**, 1522–1528.
7. Esquivel, E. V. and Murr, L. E. (2004) TEM analysis of nanoparticulates in a polar ice core, *Mater. Character.*, **52**, 15–25.
8. Fenoglio, I., Tomatis, M., Lison, D., Muller, J., Fonseca, A., Nagy, J. and Fubini, B. (2006) Reactivity of carbon-nanotubes: Free radical generation or scavenging activity?, *Free Radic. Biol. Med.*, **40**, 1227–1233.
9. Fiorito, S., Serafino, A., Andreola, F. and Bernier, P. (2006) Effects of fullerenes and single-wall carbon nanotubes on murine and human macrophages, *Carbon*, **44**, 1101–1106.
10. Huczko, A. and Lange, H. (2001) Carbon nanotubes: Experimental evidence for a null risk of skin irritation and allergy, Fullerenes, *Nanotubes and Carbon Nanostructures*, **9**, 247–250.
11. Huczko, A., Lange, H., Calko, E., Grubek-Jaworska, H. and Droszez, P. (1997) Physiological testing of carbon nanotubes: are they asbestos-like? *Full. Sci. Technol.*, **9**, 251–254.
12. Kagan, V. E., Tyurina, Y. Y., Tyurina, V. A., Konduru, N. V., Potapovich, A. I., Osipov, A. N., Kisin, E. R., Schwegler-Berry, D., Mercer, R., Castranova, V. and

Shvedova, A. A. (2006) Direct and indirect effects of single walled carbon nanotubes on RAW 264.7 macrophages: Role of iron, *Toxicol. Lett.*, **165**, 88–100.

13. Kostarelos, K., Lacerda, L., Pastorin, G., Wu, W., Wieckowski, S., Luangsivilay, J., godefroy, S., Pantarotto, D., Briand, J. P., Muller, S., Prato, M. and Bianco, A. (2007) Cellular uptake of functionalized carbon nanotubes is independent of functional group and cell type, *Nature Nanotechnol.*, **2**, 108–113.
14. Magrez, A., Kasas, S., Salicio, V., Pasquier, N., Seo, J. W., Celio, M., Catsicas, S., Scwaller, B. and Forro' L. (2006) Cellular toxicity of Carbon-based nanomaterials, *Nano Lett.*, **6**, 1121–1125.
15. Mangum, J. B., Turpin, E. A., Antao-Menezes, A., Cesta, M. F., Bermudez, E. and Bonner, J. C. (2006) Single-Walled Carbon Nanotubes (SWCNT)-Induced interstitial fibrosis in the lungs of rats is associated with increased levels of PDGF mRNA and the formation of unique carbon structures that bridge alveolar macrophages *in situ*, *Particle and Fibre Toxicology*, **3**, 15–25.
16. Manna, S. K., Sarkar, S., Barr, J., Wise, K., Barrera, E. V., Jejelowo, O., Rice-Ficht, A. C. and Ramesh, G. T. (2005) Single-walled carbon nanotubes induce oxidative stress and activate nuclear transcription factor-kappaB in human keratinocytes, *Nano Lett.*, **5**, 1676–1684.
17. Monteiro-Riviere, N. A., Nemanich, R. J., Inman, A. O., Wang, Y. Y. and Riviere, J. E. (2005) Multi-walled carbon nanotubes interactions with human epidermal keratinocytes, *Toxicol. Lett.*, **155**, 377–384.
18. Muller, J., Huaux, F., Moreau, N., Misson, P., Heilier, J. F., Delos, M., Arras, M., Fonseca, A., Nagy, J. B. and Lison, D. (2005) Respiratory toxicity of multi-wall carbon nanotubes, *Toxicol. Appl. Pharmacol.*, **207**, 221–231.
19. Murr, L. E., Bang, J. J., Esquivel, E. V., Guerrero, P. A. and Lopez, D. A. (2004) Carbon nanotubes and nanocrystal forms, and complex nanoparticle aggregates in common fuel-gas combustion sources and the ambient air, *J. Nanoparticles Res.*, **6**, 241–251.
20. Murr, L. E., Garza, K. M., Soto, K. F., Carrasco, A., Powell, T. G., Ramirez, D. A., Guerrero, P. A., Lopez, D. A. and Venzor III, J. (2005) Cytotoxicity Assessment of some carbon nanotubes and related carbon nanoparticle aggregates and the implications for anthropogenic carbon nanotubes aggregates in the environment, *Int. J. Environ. Res. Public Health*, **2**, 31–42.
21. Murr, L. E., Soto, K. F., Esquivel, E. V., Bang, J. J., Guerrero, P. A., lopez, D. A. and Ramirez, D. A., (2004) Carbon nanotubes and other fullerene-related nanocrystals in the environment: TEM study, *JOM*, **56**, 28–31.
22. Nimmagadda, A., Thurston, K., Nollert, M. U. and McFetridge, P. S. (2006) Chemical modifications of SWNT alters *in vitro* cell-SWNT interactions, *J. Biomed. Mater. Res. A*, **76**, 614–625.
23. Pulskamp, K., Diabate', S. and Drug, H. F. (2007) Carbon nanotubes show no sign of acute toxicity but induce intracellular reactive species in dependence on contaminants, *Toxicol. Lett.*, **168**, 58–74.

24. Radomski, A., Jurasz, P., Alonso-Escolano, D., Drews, M., Morando, M., Malinski, T. and Radomski, M. W. (2005) Nanoparticle-induced platelet aggregation and vascular trombosis, *Br. J. Pharmacol.*, **146**, 882–893.
25. Raja, P. M., Connolley, J., Ganesan, G. P., Ci, L., Ajayan, P. M., Nalamasu, O. and Thompson, D. M. (2007) Impact of carbon nanotubes exposure, dosage and aggregation on smooth muscle cells, *Toxicol. Lett.*, **169**, 51–63.
26. Rancan, F., Rosan, S., F., Boehm, F., Cantrell, A., Brellreich, M., Hirsch, A. and Moussa, F. (2002) *J. Photochem. Photobiol.*, B **67**, 157.
27. Salvador-Morales, C., Flahaut, E., Sim, E., Sloan, J., Green, M. L. H. and Sim, R. B. (2006). Complement activation and protein adsorption by carbon nanotubes, *Mol. Immunol.*, **43**, 193–201.
28. Sarkar, S., Sharma, C., Yog, R., Periakaruppan, A., Jejelowo, O., Thomas, R., Barrera, E. V., Rice-Ficht, A. C., Wilson, B. L. and Ramesh, G. T. (2007). Analysis of stress responsive genes induced by single-walled carbon nanotubes in BJ Foreskin cells, *J. Nanosci. Nanotechnol.*, **7**, 584–592.
29. Sato, Y., Yokoyama, A., Shibata, K., Akimoto, Y., Ogino, S., Nodasaka, Y., Kohgo, T., Tamura, K., Akasaka, T., Uo, M., Motomiya, K., Jeyadevan, B., Ishiguro, M., Hatakeyama, R., Watari, F. and Tohji, K. (2005) Influence of length on cytotoxicity of multi-walled carbon nanotubes against human acute monocytic leukaemia cell line THP-1 *in vitro* and subcutaneous tissue of rats *in vivo*, *Mol. BioSyst.*, **1**, 176–182.
30. Sayes, C. M., Liang, F., Hudson, J. L., Mendez, J., Guo, W., Beach, J. M., Moore, V. C., Doyle, C. D., West, J. L., Billups, W. E., Ausman, K. D. and Colvin, V. L. (2006) Functionalization density dependence of single-walled carbon nanotubes cytotoxicity in vitro, *Toxicol. Lett.*, **161**, 135–142.
31. Shvedova, A. A., Castranova, V., Kisin, E. R., Schwegler-Berry, D., Murray, A. R., Gandelsman, V. Z., Maynard, A. and Baron, P. (2003) Exposure to carbon nanotubes material: assessment of nanotubes cytotoxicity using human keratinocyte cells, *Toxicol. Environ. Health* A, **66**, 1909–1926.
32. Shvedova, A. A., Kisin, E. R., Mercer, R., Murray, A. R., Johnson, V. J., Potapovich, A. I., Tyurina, Y. Y., *et al.* (2005) Unusual inflammatory and fibrogenic pulmonary responses to single-walled carbon nanotubes in mice, *Am. J. Physiol. Lung Cell. Mol. Physiol.*, **289**, L698–L708.
33. Singh, R. Pantarotto, D., Lacerda, L., Pastorin, G., Klumpp, C., Prato, M., Bianco, A. and Kostarelos, K. (2006) Tissue biodistribution and blood clearance rates of intravenously administered carbon nanotubes radiotracers, *Proc. Nat. Acad. Sci. USA*, **103**, 3357–3362.
34. Tian, F., Cui, D., Schwarz, H., Gomez Estrada, G. and Kobayashi, H. (2006) Cytotoxicity of single-wall carbon nanotubes on human fibroblasts, *Toxicol. In Vitro*, **20** 1202–1212.
35. Warheit, D. B., Laurence, B. R., Reed, K. L., Roach, D. H., Reynolds, G. A. M. and Webb, T. R. (2004) Comparative pulmonary toxicity assessment of single-wall carbon nanotubes in rats, *Toxicol. Sci.*, **77**, 117–125.

36. Wick, P., Manser, P., Limbach, L. K., Dettlaff-Weglikowska, U., Krumeich, F., Roth, S., Stark, W. J. and Bruinink, A. (2007). The degree and kind of agglomeration affect carbon nanotubes cytotoxicity, *Toxicol. Lett.*, **168**, 121–131.
37. Witzman, F. A. and Monteiro-Riviere, N. A. (2006) Multi-walled carbon nanotube exposure alters protein expression in human keratinocytes, *Nanomedicine*, **2**, 158–168.
38. Zhang, L. W., Zeng, L., Barron, A. R. and Monteiro-Riviere, N. A. (2007) Biological interactions of functionalized single-wall carbon nanotubes in human epidermal keratinocytes, *Int. J. Toxicol.*, **26**, 103–113.

6

Mechanisms of Toxicity of Carbon Nanotubes

Silvana Fiorito and Annalucia Serafino

This almost exhaustive overview on the risk assessment of carbon nanotubes and on their potential beneficial use in the biomedical field, shows that the results of the experiments performed up to date on the toxicity of these C-nanoparticles are still not universal and need to be viewed as the basis for future investigations. However, some emerging concepts of nanotoxicology can be identified from these data. Toxicity of C-nanotubes depends on several different factors, some belonging to CNT properties such as: size, shape, surface characteristics, amount of the substances present in the particles preparations (carbonaceous material, graphite particles, amorphous carbon), degree of functionalisation, degree of solubilisation, length of CNTs, degree and kind of agglomeration, catalyst metal; some to the biological environment: cell line, tissue type, animal species, route of administration, type of assay used for the measurement of cell viability (in some cases it has been demonstrated that CNTs can interfere with the dye resulting in false positive and/or negative results), etc. Furthermore, the physical form of carbon, such as the molecular structure and topology has been found to be essential for assessing the toxicity of any carbonaceous material. It has been demonstrated that the degree of solubility and dispersion of CNTs can completely change the behaviour of CNTs when in contact with cells.[18] The more hydrophobic pristine CNTs have been shown to be less toxic than the oxidized CNTs. The increased toxicity of oxidized CNTs being explained because they are better dispersed in aqueous solution and therefore can reach higher concentrations of free CNT at similar weight per volume values.[1]

6.1. Cytotoxicity Mediated by Oxidative Stress

One of the main mechanisms involved in the cytotoxicity of CNTs seems to be the induction of oxidative stress in cells through the up-regulation of the pro-oxidant molecules and the down regulation of the anti-oxidant systems, as indicated by the formation of free radical species, accumulation of peroxidative products, reduction of vitamin E and total antioxidant reserves in the cells. This mechanism has been demonstrated in a study[14] in which the imbalance between the formation of peroxidative products and the reduction of antioxidants substances by skin cells in contact with SWNTs containing about 30% of iron (by weight), a redox active metal, as catalyst, has been evoked as a cause of cell toxicity. Iron overload toxicity has been generally associated with free-radical-mediated tissue damage and the development and progression of several pathological conditions including certain cancers, skin, liver and hearth disease, hormonal abnormalities and immune system dysfunctions. The cytotoxicity of some CNTs preparation, as suggested by Shvedova *et al.* could be due to iron catalyst particles that, within the SWNTs material, are encased in carbonaceous structure. This hypothesis was confirmed by the finding that an iron chelator dramatically reduced the cytotoxicity of SWNTs. The same mechanism has been demonstrated also by others[4] who analysed the interactions between two types of SWNTs (iron-rich and iron-stripped) with a macrophage cell line. They found that non purified iron-rich SWNTs were more effective in generating hydroxyl radicals as compared to purified SWNTs, and caused significant loss of intracellular low molecular weight thiols and accumulation of lipid hydroperoxides in macrophages. Other data, in agreement with the hypothesis that cytotoxicity of CNTs is based on the oxidative cell stress induced mainly by the presence of metal catalyst impurities, showed that not purified SWNTs and MWNTs, containing cobalt and molybden, induce a dose and time-dependent increase of intracellular reactive oxygen species and a decrease of the mithocondrial membrane potential either in rat macrophage cells and in human A549 lung cells line, whereas, the treatment with purified CNTs, containing a reduced metal catalyst content, had no effect on the cells.[10] A confirmation of the involvement of mechanisms of oxidative stress in the induction of cytotoxicity by CNTs, is given by another study,[13] in which the effects of SWNTs on the stress responsive genes in human immortalized skin fibroblast cells have been explored, and an increase for four of these genes has been found. The authors concluded that some of the toxic responses can partially be explained by the altered stress-related genes. The induction of increased oxidative stress and inhibition of cell proliferation in human keratinocyte cells by SWNTs, was also observed by Manna *et al.*[7] In addition, they found that SWNTs activate NF-kappaB in a dose dependent manner, and that the mechanism of activation of NF-kappaB was due to the activation of stress-related kinases. NF-kappaB are proteic factors that are involved in the control of a large number of normal cellular and organismal processes, such as, immune and inflammatory responses, developmental processes, cellular growth, and apoptosis. In most cells, NF-κB is present as a latent inactive complex in the cytoplasm. When a cell receives any of a multitude of extracellular signals, NF-kappaB rapidly enters the nucleus and activates inflammatory related-gene expression. These transcription factors are persistently active in a number of disease states, including cancer, arthritis, chronic inflammation, asthma, neurodegenerative diseases, and heart disease.

6.2. Cytotoxicity Mediated by Altered Gene Expression

The analysis of the whole genome expression in human skin fibroblast cells exposed to MWNTs revealed that multiple cellular patways were perturbed after exposure to these nanomaterials. MWNTs activated genes involved in cellular transport, metabolism, cell cycle regulation and stress response, indicative of a strong immune and inflammatory response.[3] A similar mechanism that could be responsible for cell apoptosis and decreased cellular adhesive ability has been suggested by Cui *et al.*[2] They showed that SWNTs inhibit the proliferation of human embryo kidney cells (HEK293) and decrease cell adhesion capacity, by down-regulating adhesion associated genes, and induce cell apoptosis and death by up-regulating apoptosis associated genes. They hypothesized that, after the attachment of the SWNTs to the surface of cells, a stimuli signal is transduced inside the cells and the nucleus, leading to down regulation of adhesion associated genes and corresponding adhesive proteins (laminin, fibronectin, FAK, cadherin, collagen IV and padh9), resulting in decrease of cell adhesion and causing cells to detach, float and shrink in size. At the same time, an up-regulation of apoptosis-associated genes induced by SWNTs was observed, causing cells arrest in the G1 phase, finally resulting in apoptosis. However, during this period, cells actively responded to SWNTs for self protection purposes, secreting small proteins to aggregate and wrap SWNTs into nodular structures, which isolate the cells attached by CNTs from the remaining cell mass. Surface area and surface chemistry, have been regarded as the main variables involved in the toxic effect of CNTs. Tian *et al.*[16] explored the cytotoxicity of two types of SWNTs, purified and not-purified, on human dermis fibroblast cells. Purified SWNTs, without catalytic transition metals, were found to be well dispersed and more toxic that all the other material tested. They assessed cell survival and attachment assays, immunocytochemical analysis, western blot analysis for the assessment of human fibronectin, laminin, cyclin-D3, collagen IV, beta-actin, P-cadherin, FAK (protein that play an important role in cell adhesion). Based on the observation that the expression of these proteins in the cells treated with purified SWNTs was lower as compared to normal cells, they hypothesized an interesting mechanism of action for SWNTs cytotoxicity. SWNTs would activate extracellular matrix (ECM) protein signals, and thus, the cell starts changing the cytoskeleton. Afterwards, a displacement of internal organelles and a deformation of the cell membrane take place. Then the decrease of fibronectin, laminin, P-cadherin, FAK, cyclin D3 and collagen IV expression levels induces cell detachment, as shown by the observed morphological change in shape and adhesion features (Fig. 1), with consequent cell apoptosis/necrosis. The authors suggested

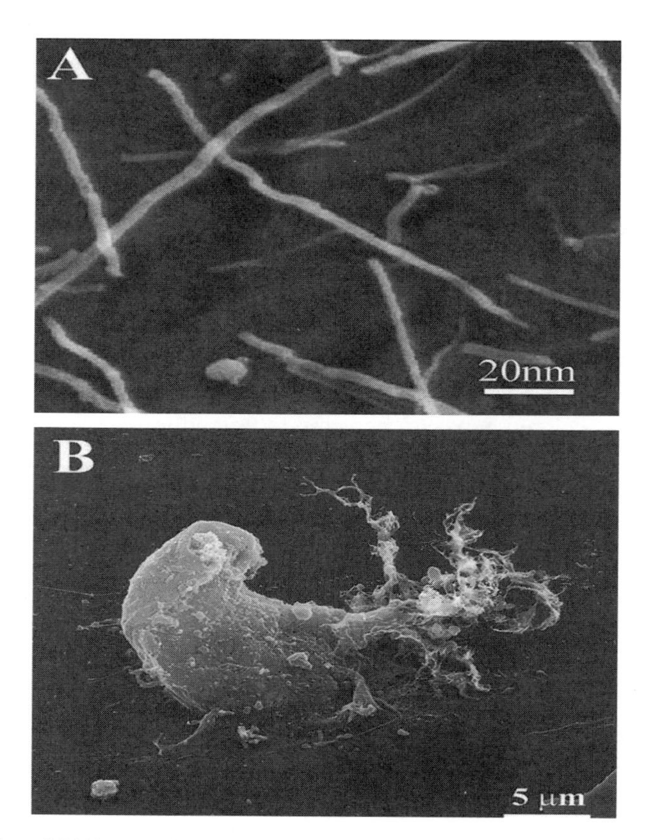

Fig. 1. Effect of SWCNTs on human fibroblast cells: (A) scanning electron microscopy image of dispersed SWCNTs over the substrate, they have the sharpest shape, among the five nanomaterials, due to a rather large aspectratio; (B) change in cell spreading seen on samples treated with SWCNTs. From, Tian F. *et al.*, *Toxicology in vitro* (2006), **20**, 1202–1212.

that the surface chemistry has the leading role in the cytotoxicity mechanism. Dispersed purified SWNTs and MWNTs, due to a change in surface chemistry that modifies both the surface area and the aggregation, were shown to be more harmful than not purified SWNTs and MWNTs.

In order to explain the mechanisms underlying the formation of interstitial fibrotic lesions and granuloma in lung tissue after pharyngeal aspiration or instillation, changes in the expression of genes encoding known profibrogenic mediators *in vivo*, following SWNTs exposure were investigated[6] and gene expression patterns were correlated with the formation of fibrotic lesions. The observed induction of increased lung mRNA levels encoding pro-fibrogenic factors (PDGF ligands), at day 1 after SWNTs exposure, suggested that PDGF could play a role in the formation of the fibrotic lesions. Moreover, the fact that only SWNTs stimulated fibrotic responses, as compared to carbon black particles, that possessed similar specific surface areas but did not contain contaminating metals, indicates that the fibrogenic activity of SWNTs is likely due to either differences in shape or elemental composition. The SWNTs used in this study were synthesized by chemical vapour deposition using cobalt and molybdenum as catalysts, and contained 2.6% and 1.7% respectively of these metals that could contribute to the surface reactivity of SWNTs. Thus, it has been proposed that low levels of contaminating metals, coupled with high surface area, could determine the toxicity and fibrogenic potential of CNTs. This is in agreement with other reports showing SWNTs-induced lung interstitial fibrosis.[5,15,17] In these papers the use of SWNTs containing residual metal catalysts (nichel, molybdenum and iron) was described, and, at least for iron, it was demonstrated that the metal could be considered responsible for the cytotoxicity of the CNTs.

6.3. Cytotoxicity Mediated by Geometrical and Mechanical Factors

A very surprising observation has been made by two research groups who, working separately, both described the formation of carbon bridge structures between either alveolar macrophages *in situ*[6] and platelets.[11] This phenomenon could be due to the structural characteristics of CNTs and/or to the similarity existing between carbon nano-structures and the nano-filaments that link immune cells each other when they are activated. These immune cell nanotubes are derived in part from cell surface membrane and are able to connect various cells simultaneously, thereby establishing complex communication networks between immune cells.[8] This interesting similarity existing between cell nanostructures and carbon nanostructures raises many important new questions about how carbon nanotubes could interact with cells, and whether it would be possible, in a next future, to use this capacity to make them useful therapeuthic or diagnostic tools. The mechanism behind the formation of cell bridges structures is still unclear but it could explain some of the interactions between CNTs and cells, i.e. the induction of platelet aggregation in absence of platelet degranulation and activation, as described by Radomsky,[11] or the inhibition of macrophage functions as suggested by Mangum.[6] Moreover, it has been found[9] that SWNTs constitute a new class of universal K^+ channel inhibitors exploiting rather elementary principles of blocking. The authors suggest that the geometry and not the chemical nature of the molecule must be the primary factor governing blockade mechanism. In fact, they found that the extent of the block varied among channel types, but in all cases blocking was stronger with small diameter nanotubes and was suppressed with large diameter nanostructures (MWNTs did not exert any block). SWNTs probably hamper channel function, by fitting into the pore and thus either hindering ion movement or alternatively preventing further conformational steps. In this study another previously unknown novel property of these nanomaterials has been unveiled, that represents a further step forward in the effort to utilize them for biological purposes. Based on the finding that the toxicity of CNT-bundles is lower than that of CNT-agglomerates,[18] it has been hypothesized that one of the cytotoxicity mechanisms by which CNT-agglomerates affect cell viability is that these carbon-structures, bigger, stiffer and more solid (similar to asbestos) than the bundles in CNT-bundles fraction, induce a cytotoxic response of the cell comparable to asbestos. Thus, a mechanical effect due to the size and to the degree of dispersion of the CNTs, seems to be another important factor leading to cytotoxic effects.

6.4. Cytotoxicity Mediated by the Binding of CNTs to Plasma Proteins

The activation of the human serum complement system by different types of CNTs, including SWNTs and DWNTs was shown by Salvador-Morales, *et al.*[12] Activation of the human serum complement system via the classical pathways, takes place when the protein C1q, the recognition subunit of the C1 complex, binds to complement activators. As a consequence of complement activation, fragments of various complement components are generated, that lead to the generation of an inflammatory response and also to the formation of granulomas at a later stage. In order to elucidate the mechanism of activation and to analyse the interaction of complement proteins with CNTs the authors studied the binding of C1q to carbon nanotubes by using Western blotting. When the CNTs were exposed to human serum and plasma, only a few proteins bind directly and in large quantity to CNTs: C1q, fibrinogen and apolipoprotein A1. Thus, the direct binding of the protein C1q to CNT, might be responsible for the complement system activation and consequent generation of inflammatory peptides. The activation of human complement induced by CNTs might be diminished or eliminated by alteration of surface chemistry. The authors hypothesize that, in a next future, it will be possible to modulate this effect through a variation in surface charge, that, for example, could promote binding of factors down regulators of complement activation.

To date, several mechanisms of action have been suggested for explaining CNTs toxicity. Results are not completely consistent each other, because of different materials, different cell types and different experimental conditions used. Nevertheless, it seems clear that the following factors are the more important in determining the cytotoxic potential of these nanostructures: (i) the presence of impurities into the sample, (ii) the degree of dispersion, (iii) the surface properties, (iv) the shape of the nanotubes.

Acknowledgement

We are grateful to Martino Tony Miele for the technical expertise and assistance provided.

Bibliography

1. Bottini, M., Bruckner, S., Nika, K., Bottini, N., Bellucci, S., Magrini, A., Bergamaschi, A. and Mustelin, T. (2006) Multi-walled carbon nanotubes induce T lymphocyte apoptosis, *Toxicol. Lett.*, **160**, 121–126.
2. Cui, D., Tian, F., Ozkan, C. S., Wang, M. and Gao, H. (2005) Effect of single wall carbon nanotubes on human HEK293 cells, *Toxicol. Lett.*, **155**, 73–85.
3. Ding, L., Stilwell, J., Zhang, T., Elboudwarej, O., Jiang, H., Selegue, J. P., Cooke, P. A., Gray, J. W. and Chen, F. F. (2005) Molecular characterization of the cytotoxic mechanism of multiwall carbon nanotubes and nano-onions on human skin fibroblast, *Nano Lett.*, **5**, 2448–2464.

4. Kagan, V. E., Tyurina, Y. Y., Tyurina, V. A., Konduru, N. V., Potapovich, A. I., Osipov, A. N., Kisin, E. R., Schwegler-Berry, D., Mercer, R., Castranova, V. and Shvedova, A. A. (2006) Direct and indirect effects of single walled carbon nanotubes on RAW 264.7 macrophages: Role of iron, *Toxicol. Lett.*, **165**, 88–100.
5. Lam, C., James, J. T., McCluskey, R. and Hunter, R. L. (2004) Pulmonary toxicity of single-wall carbon nanotubes in mice 7 and 90 days after intratracheal instillation, *Toxicol. Sci.* **77**, 126–134.
6. Mangum, J. B., Turpin, E. A., Antao-Menezes, A., Cesta, M. F., Bermudez, E. and Bonner, J. C. (2006) Single-Walled Carbon Nanotubes (SWCNT)-induced interstitial fibrosis in the lungs of rats is associated with increased levels of PDGF mRNA and the formation of unique carbon structures that bridge alveolar macrophages in situ, *Particle and Fibre Toxicology*, **3**, 15–25.
7. Manna, S. K., Sarkar, S., Barr, J., Wise, K., Barrera, E. V., Jejelowo, O., Rice-Ficht, A. C. and Ramesh, G. T. (2005) Single-walled carbon nanotubes induce oxidative stress and activate nuclear transcription factor-kappaB in human keratinocytes, *Nano Lett.*, **5**, 1676–1684.
8. Önfelt, B., Nedvetzki, S., Yanagyi, K. and Davis, D. M. (2004) Cutting edge: Membrane nanotubes connect immune cells, *J. Immunol.*, **173**, 1511–1513.
9. Park, K. H., Chhowallas, M., Iqbal, Z. and Sesti, F. (2003) Single-walled carbon nanotubes are a new class of ion channel blockers, *J. Biol. Chem.*, **278**, 50212–50216.
10. Pulskamp, K., Diabate', S. and Drug, H. F. (2007) Carbon nanotubes show no sign of acute toxicity but induce intracellular reactive species in dependence on contaminants, *Toxicol. Lett.*, **168**, 58–74.
11. Radomski, A., Jurasz, P., Alonso-Escolano, D., Drews, M., Morando, M., Malinski, T. and Radomski, M. W. (2005) Nanoparticle-induced platelet aggregation and vascular trombosis, *Br. J. Pharmacol.*, **146**, 882–893.
12. Salvador-Morales, C., Flahaut, E., Sim, E., Sloan, J., Green, M. L. H. and Sim, R. B. (2006) Complement activation and protein adsorption by carbon nanotubes, *Mol. Immunol.*, **43**, 193–201.
13. Sarkar, S., Sharma, C., Yog, R., Periakaruppan, A., Jejelowo, O., Thomas, R., Barrera, E. V., Rice-Ficht, A. C., Wilson, B. L. and Ramesh, G. T. (2007) Analysis of stress responsive genes induced by single-walled carbon nanotubes in BJ foreskin cells, *J. Nanosci. Nanotechnol.*, **7**, 584–592.
14. Shvedova, A. A., Castranova, V., Kisin, E. R., Schwegler-Berry, D., Murray, A. R., Gandelsman, V. Z., Maynard, A. and Baron, P. (2003) Exposure to carbon nanotubes material: Assessment of nanotubes cytotoxicity using human keratinocyte cells, *Toxicol. Environ. Health A*, **66**, 1909–1926.
15. Shvedova, A. A., Kisin, E. R., Mercer, R., Murray, A. R., Johnson, V. J., Potapovich, A. I., Tyurina, Y. Y., *et al.* (2005) Unusual inflammatory and fibrogenic pulmonary responses to single-walled carbon nanotubes in mice, *Am. J. Physiol. Lung Cell. Mol. Physiol.*, **289**, L698–L708.

16. Tian, F., Cui, D., Schwarz, H., Gomez Estrada, G. and Kobayashi, H. (2006) Cytotoxicity of single-wall carbon nanotubes on human fibroblasts, *Toxicol. In Vitro*, **20**, 1202–1212.
17. Warheit, D. B., Laurence, B. R., Reed, K. L., Roach, D. H., Reynolds, G. A. M. and Webb, T. R. (2004) Comparative pulmonary toxicity assessment of single-wall carbon nanotubes in rats, *Toxicol. Sci.*, **77**, 117–125.
18. Wick, P., Manser, P., Limbach, L. K., Dettlaff-Weglikowska, U., Krumeich, F., Roth, S., Stark, W. J. and Bruinink, A. (2007) The degree and kind of agglomeration affect carbon nanotubes cytotoxicity, *Toxicol. Lett.*, **168**, 121–131.

7

Conclusions

The new world of health started since nanotechnology has demonstrated that it will be possible, in a next future, to identify diseases at the earliest stages — at the cellular level — and to deliver therapeutic drugs (that are only activated when they reach their target) directly to the disease site reducing their toxicity. This revolutionary change in the dimensions of our "medical approach" is going to take us closer to our inner natural dimensions and to make us to appreciate all those things that belong to the nano-sized world.

The benefits of this emerging research field are huge and even if therefore some issues concerning nanomaterials toxicity are still to be elucidated, the possible risks deriving from nanomaterials application appear to be a minor problem.

In the last few years, the development of nanotechnology has included many funds from different countries all around the world for studying environmental, health and safety issues relating to nanoscale materials and understanding the risk-benefit assessments. Moreover, the public's acceptance of nanotechnology plays an increasingly important role in determining the ultimate impact nanotechnology will have across society. The studies in progress on the future biomedical applications of Carbon nanotubes lead to the development of new tools for nanodiagnostic, including biosensors and medical imaging, regenerative medicine, scaffolds and orthopaedic and neuronal devices, drug delivery and release systems for therapeutic purposes. But till now, speculations about C-nanoparticles toxicity are still inconclusive.

Before any investigation is done and conclusions about their biocompatibility drawn, we need to know how the particles have been synthesized and dispersed, and the amount of metal catalysts and graphite present in the preparation, their surface modifications and functionalization, their surface chemistry (coating) and chemically reactive sites (free radicals). Furthermore, additional considerations for assessing safety of C-nanoparticles include a careful selection of appropriate doses/concentrations related to the cell types and tissue species.

Current studies have assessed toxicity of nanotubes when placed in contact with living systems by an artificial mechanism: implantation, instillation in small animals, incubation with cells *in vitro*, using dose levels that probably do not reflect what would be routine handling and use of nanomaterials.

On one hand, little is currently known about realistic exposure levels, especially lung exposure. The capacity of nanotubes to form persistent agglomerates can make the size of the inhaled particles much greater than the individual tube size. On the other hand, nothing is known about the exposure of nanomaterials to animals or humans by direct contact, and how the natural defence systems, that may discriminate, segregate, modify or even eliminate nanomaterials throughout the body, could behave when accidentally exposed to these materials.

Another important issue is that most of the published toxicologic studies have shown only short term results, but nothing is known about the long term effects of CNTs. Nanomaterials, once inhaled or penetrated into the bloodstream, may interfere with endogenous metabolic or signal transduction pathways, and perturb cellular biochemical functions in more subtle way that cannot be apparent in short-term toxicity assays.

In addition, another limitation to the interpretation of these studies is due to the polyedric behaviour of carbon nanotubes towards the biological environment. They can be regarded, at the same time, as hazardous as well as beneficial materials, depending on the conditions under which their toxicity is evaluated. There are so many factors influencing their biocompatibility and/or toxicity hence it is senseless to speculate about toxicity of carbon nanotubes in general. The evaluation of the health hazard towards humans and animals of these nanostructures has to be focused on the specific purpose for which they will be used. The same nanotubes can have beneficial and useful effects in some conditions (i.e. free radical scavenging activity) or be hazardous to humans and animals in other situations.

Furthermore, the toxicity of C-nanotubes needs to be understood in the framework of the material characterization. If scientists do not understand the material from a physical and chemical perspective, they cannot interpret exposure or toxicity measurements.

We would like to underline that an interdisciplinary team approach is imperative for nanotoxicology research to reach an appropriate risk assessment in terms of toxicity of C-nanoderivatives to the biological environment in order to quiet their useless demonization and allow for their use as beneficial tools in the more various fields.

Silvana Fiorito

Index

3D architectures, 70

acetabular cup prosthesis, 81
acid treatment, 114
activation-inflammation injury, 116
actual, 27
actuators, 19, 69
acute
  inflammation, 111
  lung toxicity, 111
  toxicity, 107, 114
adherent mammalian cells, 106
adhesion associated genes, 129
adhesive proteins, 129
agglomeration, 121, 127
aggregated MWNTs, 108
aging population, 94
Aharonov-Bohm effect, 20
airways, 111
alginate nano-composite gel, 71
allergies, 109
alumina, 90
alveolar
  macrophages, 112, 131
  region, 112
alveolitis, 111
amino acids, 47
amorphous carbon, 5, 127
amperometric biosensors, 22
Amphotericin B (AmB), 48
animal species, 127
anthropogenic nanoparticulates, 108
anti-oxidant systems, 127
antibody responses, 49
antioxidant, 113
antioxidants substances, 128
apoptosis, 116, 129
  associated genes, 129
  necrosis, 121
applications, 27
aqueous
  solution, 127
  suspension, 113
armchair, 7, 25
  nanotubes, 9
artificial
  bone, 79
  hips and knees, 88
  implant, 61
  muscles, 69
  nanostructures, 67
  replacements, 71
as-produced nanotubes, 5
asbestos, 119, 132
aspect ratio, 114
asthmatic patients, 109

B and T lymphocytes, 119
bacteria cells, 106
bio
  active and resorbable biomaterials, 63
  conductive, 63
  matrices, 93
  mimetic sensors, 69
  mimetic tissue surfaces, 77
  resorbable polymers, 63
bioactive
  peptides, 47
  proteins, 47
biocompatibility, 62
  of CNTs, 121
biocompatible matrix, 70
biodegradable
  biomaterials, 62
  conductive material, 80
biodistribution, 43
biologic substitutes, 43
biological
  applications, 93
  environment, 105, 127
  inertness, 62
  macromolecules, 93
  toxicological behavior, 93
biomarkers, 36
biomaterials, 92
biomechanical response, 94
biomedical
  applications, 36
  field, 127
biomolecules, 69

biosensor devices, 36
biosensors, 36, 93
black
  carbon aggregates, 109
  soot, 5
blood circulation, 106
bone, 108
  formation, 43
  fractures, 78
  grafts, 79
  ongrowth, 82
  tissue, 77
brain, 108
  damage, 73
  probes, 72
bronchoalveolar lavage, 110
bronchoalver space, 110
brought, 52
bucky ball, 1
bundles, 12

C_{60}, 1, 4
C_{60}-fullerenes, 115
C_{70}, 4
C1q, 133
cancer, 128
  biomarkers, 38
  detection, 52
  markers, 52
  nanotechnology, 52
  therapy, 51
carbon
  atoms, 3, 5
  black particles, 131
  bonds, 20
  bridge structures, 131
  clusters, 1
  cylinders, 1
  fibers, 82
  filaments, 5
  nanofiber (CNF) reinforced PEEK composite, 89
  nanoforms, 109
  nanostructures, 131, 132
  nanotube actuators, 30
  nanotubes, 19
  soot, 1
carbonaceous material, 119, 127
carbonyl, 114
carboxyl, 114
carboxylic acid groups, 88
cardiovascular effects, 121
carotid artery thrombosis, 120
carrier systems, 49
cartilage, 94
catalysis, 23
catalyst, 5, 116, 119, 121, 128
  metal, 127
catalytic chemical vapor deposition, 3
cell
  adhesion, 129
  apoptosis, 118, 129
  apoptosis/necrosis, 130
  behaviour, 43
  bridges structures, 132
  culture, 113
  cycle, 119
  cycle regulation, 129
  death, 116
  growth, 93, 121
  injury effects, 111
  line, 121, 127
  morphology, 119
  nanostructures, 132
  proliferation, 43, 128
  regeneration, 77
  responses, 105
  signalling, 113
  surface membrane, 131
  therapy, 43
  uptake, 113
  vector *T*, 11
  viability, 113, 114, 119, 127
cells arrest, 129
cellular
  adhesion ability, 118
  adhesive ability, 129
  growth, 129
  imaging, 36
  transport, 119, 129
ceramic-CNT, 90
chemical
  and biological sensing, 36
  catalytic vapor desposition method, 5
  properties, 20
  sensors, 21
  surface treatment, 114
  vapour deposition, 131
Chemical Vapour Deposition technique (CVD), 115

chemistry of the surface, 70
chiral, 7
  conducting nanotubes (nanocoils), 29
  conductivity, 29
  nanotube, 9
chirality, 12, 19
chrysotile asbestos, 109
clearance half-life, 106
clearance of CNTs, 56
CNT
  agglomerates, 119, 132
  aggregates, 109
  alumina composites, 91
  based sensors, 37
  bundles, 119, 132
  dispersion, 119
  enhanced bone cement, 90
  pellet, 119
  polymer composites, 88
  row material, 119
coating, 67, 79, 92
cobalt, 115, 128, 131
collaret, 5
combustion products, 108
complement
  activation, 117
  cascade, 116
  proteins, 133
  system activation, 133
composite material, 27
composites, 19, 67
compressive strengths, 26
contaminating metals, 131
could sense, 52
cytoskeleton, 130
cytotoxic
  response, 109, 118, 121
cytotoxicity, 106
  mechanism, 131

degeneration process, 92
degree
  of dispersion, 119, 132
  of functionalisation, 121, 127
  of solubilisation, 127
  of solubility, 127
  of solubilization, 121
dental implants, 68
Deoxyribonucleic acid (DNA) biosensors, 40
dermal toxicity, 113
dermatological testing, 113
diagnostic
  approach, 36
  field, 46
diameter, 12
dispersed carbon nanomaterials, 121
dispersion of CNTs, 127
  properties, 105
distance, 2
DNA transfection, 50
double-walled-nanotubes, 116
drug delivery, 36, 51, 93
dye, 114, 127

elastic properties, 20
elasto-electronics, 29
electric
  arc, 3
  arc chamber, 4
  arc discharge, 4
  arc method, 1
  arc technique, 115
  chamber, 5
  properties, 20
electrical
  conductance, 21
  conductivity, 72
electrically excitable tissues, 75
electrodes, 22
electromechanical-tweezers, 28
electron
  charge transfer, 21
  diffraction, 14
  microscopy analysis, 108
  microscopy technique, 1
electronic
  and photonic devices, 23
  properties, 12, 20
  states, 13
electrostatic interactions, 50
embryo kidney cells, 106, 129
endo-prosthesis, 90
energy storage, 14, 19
engineered tissue, 43
environmental
  exposure, 105
  health effects, 110
epidermal keratinocytes, 113
epithelioid granulomas, 111

erythropoietin (EPO), 48
European Market, 61
excretion, 71
extra-cellular scaffold, 74
extracellular matrix (ECM) protein
  signals, 130
eyes irritation, 113

Fermi level, 13
FET device, 38
fibroblast
  cells, 80, 129
  cell cultures, 118
  cell line, 121
fibrogenic
  activity, 131
  potential, 112, 131
fibrosis, 111
fibrotic reactions, 111
field
  effect transistor (FET), 38
  emission, 14
  emission display, 28
  emitters, 19
flow sensors, 21
fluorescent probe, 47
folic acid, 53
foreign
  body granuloma, 118
  body materials, 105
fracture and fatigue fracture, 90
free radical species, 127
free radicals, 113
  scavengers, 113
fullerene, 1
fullerenic polyhedra, 108
functional group, 108
functionalisation, 47, 92, 93
functionalised
  CNTs, 50, 114
  MWNTs, 89
  PEEK, 89
  SWNTs, 78, 108
fungal cells, 106

gamma scintigraphy, 42, 108
gas break-down sensors, 19
gene, 119
  delivery system, 51
  expression, 51
  silencing, 51
  therapy, 51
  transfection, 47, 49
genetic
  diseases, 40
  vaccination, 51
genome expression, 119, 129
geometrical structures, 12
glial-scar forming astrocytes, 73
glucose biosensors, 37
granuloma, 111, 131
graphene, 9, 20
  sheet, 9
graphite, 20, 105, 115, 116
  electrodes, 4
  particles, 115, 116, 127
graphitic
  layers, 5
  structures, 1, 5
graphitization of the nanotubes, 5
guinea pigs, 110

health effects, 105
health-related issues, 92
heart, 108
helicity, 19
high
  resolution transmission microscopy, 1, 14
  temperature methods, 3
histopathology, 112
hollow nanotubes, 90
host tissues, 93
human
  and animal health, 105
  cell types, 106
  cells, 116, 119
  cells growth, 118
  complement, 117
  epidermal cells, 113
  fibroblast cells, 121
  keratinocyte cells, 128
  lung cells, 114
  monocytes-derived-macrophages (MDMs), 115
  platelet, 120
  serum complement system, 132
hybrid properties, 20
hybridization, 20
hydrocarbons, 3

hydrophobic pristine CNTs, 127
hydroxyl
  groups, 114
  or superoxide radicals, 113
  radicals, 114, 128
hyperthermia, 42

IL-1beta, 112
imaging
  agents, 42
  and tissue engineering, 36
immune
  and inflammatory responses, 128, 129
  cells, 132
immunization, 48
immunocytochemical analysis, 129
immunosensors, 38
implantable
  medical devices, 62
  sensors, 42
implanted cells, 43
implants, 91
indoor environments, 109
inflammation, 110
inflammatory
  peptides, 133
  potential, 106
  related-gene, 129
  response, 115, 118, 133
inhalation, 110
injection, 71
intercellular structures, 112
interface properties, 92
interfacial bonding, 87
interleukine 8, 113
interstitial
  fibrotic lesions, 112, 131
  inflammation, 111
intertube, 2
intestine, 108
intra-corporeal medical devices, 63
intracellular transporters, 47
intrapharyngeal instillation, 121
intratracheal instillation, 110
intravenous administration, 108
intrinsic near-infrared fluorescence, 106
in vitro diagnosis, 46
ion movement, 132
ionization CNT sensor, 22
iron, 113, 128, 131
  catalyst particles, 128
  chelator, 113, 114, 128
  rich SWNTs, 114
  stripped SWNTs, 114

joint
  implants, 90
  replacement, 82

K^+ channel inhibitors, 132
keratinocyte cells, 113
kidneys, 108
kinetics of single-walled carbon
  nanotubes, 106
knee implants, 90

laser ablation, 3
length, 118
  of CNT, 121, 127
leukemia cells, 106
ligaments, 94
light emission, 14
lipid hydroperoxides, 128
lipid peroxidation, 114
liver, 107
load transfer, 88
lubrication, 23
lung, 108
  histophatology, 111
  lesions, 111
  mRNA levels, 131
  tissue, 131
  tumor cells, 106

macrophage
  cells, 115
  encapsulation, 71
macrophages, 106, 119
magnetic
  particles, 5
  resonance spectroscopy, 5
magnetic properties, 20
mammalian
  cell suspension, 106
  cells, 106
mammals tissues, 106
markers
  of inflammation, 111
  of oxidative insults, 121
material type, 121

mechanical
  behavior, 20
  properties, 20
  response, 27
  stretching, 29
mechanics, 14
medical
  devices, 92
  imaging, 93
  implants, 61
meniscus, 94
mesothelioma cell, 106, 119
metabolic alterations, 116
metabolism, 119, 129
metal, 12, 114
  catalysts, 105, 131
  CNT, 90
metallic, 20
micro-Raman, 14
micron-sized materials, 105
microns, 2
mineral species, 108
mitochondrial membrane potential, 114, 128
moderate inflammation, 111
molecular
  bridges, 121
  structure, 5, 127
molybden, 128
molybdenum, 131
monoclonal antibodies, 42
monocytic cells, 118
multi-walled carbon nanotube (MWNT), 1, 19
multifocal granulomas, 111
multitesting, 46
murine macrophages, 106
muscle, 94, 108
musculoskeletal
  devices, 63
  system, 63
MWNT geometry, 24
MWNT/poly-L-lactide (PLLA) composite, 80
MWNTs, 108, 114, 129

nano
  composites, 69
  crystalline alumina-SWNT nano-composites, 91
  electronics, 14
  -filaments, 131
  filler, 70
  junctions, 14
  patterns, 70
  phase fibers, 78
  robots, 69
  scale integrated circuit, 75
  sized probes, 72
nanocoil, 29
nanoelectronics, 19
nanofibers, 67
nanomaterials, 105
nanomedicine, 35
nanometers, 2
nanoparticulates, 108
nanoribbons, 26
nanorope or bundle, 24
nanostructures, 19
nanosystems, 46
nanotube
  axis, 13
  bundles, 19
  field-effect transistor, 14
  functionalization, 21
  tips, 28
necrosis, 111, 118
negative surface charges, 78
nerve regeneration, 72
nervous system, 71
neural
  cell functions, 73
  electrophysiological imaging, 75
  probes, 68
  prostheses, 72
  signal transmission, 76
  stem cells, 75
  tissue, 72
neurite
  branching, 74
  outgrowth, 74
neuron growth, 43, 73
neuronal
  devices, 72
  networks, 75
  processors, 77
  signals, 75
neurons, 71
neutral surface charge, 79
neutrophil infiltration, 118

NF-kappaB, 113, 128
nickel, 115
NIR
  fluorescence, 42
  laser radiation, 42
not purified SWNTs, 114

opened SWNTs, 116
optical
  detection, 43
  properties, 20
organ biodistribution, 106
oropharyngeal aspiration, 112
orthopaedic
  devices, 64
  diseases, 63
osteoblast
  adhesion, 77
  proliferation, 43
osteointegration, 77
outdoor
  air, 108
  environment, 108
oxidative
  cell stress, 128
  stress, 127
oxidative stress, 113, 114
oxidized CNTs, 114, 127
oxydation, 113
oxygen species, 113

pancreas, 108
peapods, 15
PEEK matrix, 89
peribronchial inflammation, 111
perinuclear region, 108
peroxidative products, 113, 128
pharmacokinetic, 46, 53
pharyngeal aspiration, 110, 131
phenotypic measurements, 118
phototherapy, 42
physico-chemical behaviour, 105
platelet, 120, 131
  aggregation, 120, 132
  degranulation, 132
  -platelet interactions, 121
Poisson ratio, 25
poly
  carbonate-urethane, 78
  methyl-meth-acrylate, 90
polymer
  composites, 28
  matrix, 78
potential exposure routes, 105
pristine
  CNTs, 114
  hydrophobic and oxidized MWNTs, 114
pro-fibrogenic factors (TGF-beta), 112
pro-oxidant molecules, 127
profibrogenic mediators, 131
programmed cell death, 114
proinflammatory cytokines, 111, 112, 114, 118
properties, 20
protein C1q, 133
proteins, 113
pulmonary
  effects, 111
  function, 110
  toxicity, 110
purified
  CNTs, 114
  MWNTs, 113
  nanomaterials, 121
  SWNTs, 116, 121, 129

quantum conductance behavior, 14
quartz, 111

rabbits, 113
radical
  scavengers and antioxidants, 89
  scavenging capacity, 113
radio-labelled SWNTs, 53
radioactivity tracing, 108
radiotracers, 42
rat
  aortic smooth muscle cells, 121
  macrophage cells, 128
rats, 118
reactive
  inflammatory response, 116
  oxygen species, 114, 128
regenerative medicine, 36, 93
reinforced cement, 90
reinforcement, 92
reinforcing component, 87
release systems, 93
renal excretion, 108

respiratory diseases, 110
reticuloendothelial system organs, 108
risk assessment, 127
route of administration, 121, 127

scaffold, 67, 92, 94
scanning electron microscopy, 5
semiconducting, 20
  SWNT, 108
semiconductors, 12
sensing devices, 22
sensors, 14, 19
shape, 2, 127
sidewall functionalization, 15
sidewall functionalized SWNT, 118
single stranded DNA (ssDNA), 50
single-walled carbon nanotube (SWNT), 1, 19
size, 2, 127
  of the particles, 121
skin, 108
  cells, 106, 113
  fibroblast cells, 129
  fibroblasts, 119
  irritation, 113
  sensitivity, 113
  toxicity, 110
solar beam, 3
soot, 110
spinal cord, 108
spleen, 108
stem cells, 73
stiffness of CNT, 87
STM spectroscopy, 14
strength, 78
stress
  genes, 119
  related kinases, 113, 128
  response, 119, 129
  responsive genes, 119, 128
  transfer, 87
structural
  composites, 93
  implants, 94
structured surfaces, 67
subcutaneous
  administration, 110
  tissue, 118
sulfonated PEEK, 89
surface, 94
  area, 121, 129, 131
  characteristics, 127
  charge, 121, 133
  chemistry, 114, 121, 129, 131
  coating, 105
  energy, 73, 78
  interaction, 93
  reactivity, 131
  roughness, 78
surfactant stabilized SWNTs, 118
susceptibility, 20
SWCNTs biocompatibility, 118
SWNT-FET devices, 40
SWNT-reinforced silk nanofibers, 80
symmetry structures, 7

T and B lymphocytes, 106
T cells, 114
targeted
  cancer therapy, 53
  drug delivery, 47
targeting, 46
technological applications, 23
tendons, 94
tensile flexibility, 78
therapeutic
  molecules, 47
  purposes, 93
  tool, 36
thermal
  ablation therapy, 51
  and vibrational, 20
thermoplastic polymers, 88
thermoset polymers, 88
time, 82
tissue
  damage, 128
  engineering, 43
  growth scaffolds, 78
  interaction, 94
  regeneration, 43, 94
  scaffolds, 43
  type, 127
TNF-alfa, 112
toxic responses, 128
toxicity, 105
transcription factors, 129
transfection agents, 51
transmission electron microscopy (TEM), 113, 114
transplanted cells, 43
transport properties, 29

trauma, 63, 94
treatment of osteoporosis, 79
tribological properties, 89
tubes, 25
tumor cells, 53
"tumor-directed" therapy, 56

ultrastructural alterations, 116
undecorated SWNTs, 118
uptake, 115
urine excretion, 108

vaccination peptides delivery, 48
vaccines antigens delivery, 49
van der Waals interaction, 3, 12
van Hove singularities, 13
vaporization of graphite, 1
vascular stents, 69
vector
  P, 8
  T, 11
  model, 8
vibrational properties, 14
viral-peptide-CNTs, 48
viruses, 40
vitamin E, 113, 128

water soluble
  carbon nanotubes, 118
  SWNTs, 118
wear
  rate, 89
  resistance, 90, 94
well-dispersed CNTs, 119
western blot analysis, 129
workers, 113
world market, 61

X-ray contrast agents, 42

yeast cells, 106
Young's modulus, 20

Zeeman splitting, 20
zigzag, 7
  nanotubes, 9